Tamiris Augusto Marinho

HBV infection in waste pickers in Goiânia-GO

Tamiris Augusto Marinho

HBV infection in waste pickers in Goiânia-GO

An epidemiological investigation

ScienciaScripts

Imprint
Any brand names and product names mentioned in this book are subject to trademark, brand or patent protection and are trademarks or registered trademarks of their respective holders. The use of brand names, product names, common names, trade names, product descriptions etc. even without a particular marking in this work is in no way to be construed to mean that such names may be regarded as unrestricted in respect of trademark and brand protection legislation and could thus be used by anyone.

Cover image: www.ingimage.com

This book is a translation from the original published under ISBN 978-3-330-75392-1.

Publisher:
Sciencia Scripts
is a trademark of
Dodo Books Indian Ocean Ltd. and OmniScriptum S.R.L publishing group

120 High Road, East Finchley, London, N2 9ED, United Kingdom
Str. Armeneasca 28/1, office 1, Chisinau MD-2012, Republic of Moldova, Europe
Printed at: see last page
ISBN: 978-620-7-89609-7

Table of contents:

DEDICATORY

ACKNOWLEDGMENTS

Giving thanks for the realization of this dream is a very special moment, after all, I have received so many divine blessings along the way... always mediated by wonderful friends and incredible learning opportunities. My gratitude is immeasurable, after all, sharing this victory is the best thanks I could give to all of you:

To God for granting me another life, and with it this unique moment of growth, thank you my Father for your infinite mercy;

To my advisor, Prof. Dr. Regina Maria Bringel Martins, for guiding me in my scientific initiation during four years of undergraduate studies and offering me the opportunity to enter the master's program, sparing no effort for the success of our study. Thank you for so many opportunities, you will always be my benchmark for professional competence;

To Prof. Dr. Carmen Luci Rodrigues Lopes, for being such a partner and friend. She was an admirable person, always willing to help with affection and delicacy, characteristics of her kind personality;

To Prof. Dr. Sheila Araújo Teles, my teacher during my undergraduate and master's studies, a great example of a professional, as a teacher, researcher and nurse, thank you especially for your support in the statistical analysis of the data in this study;

To Prof. Dr. Megmar Aparecida dos Santos Carneiro, who was always present in the Virology Laboratory and taught me great lessons in life, an example of a teacher, researcher, master and woman;

To Prof. Dr. Márcia Alves Dias de Matos, for always being so competent... I'm at a loss for words to thank you for everything I've learned from you, thank you for teaching me about molecular phylogeny, for being so willing and helpful with all my questions. You are a great example, a strong, hard-working and humble woman who has overcome all obstacles and is now a winner;

To professors Sandra Brunini, Ruth Minamisava, Divina Cardoso, Fabíola Fiaccadori, Menira Souza, André Kipnis and Joao Bosco, for teaching me valuable knowledge during my postgraduate studies;

To the entire Virology Laboratory team: Nádia Rúbia, Aline Garcia, Ágabo, Aline Pereira, Renata, Andréia, Dulce, Kamilla, Láiza, Laura, Lyriane, Fernanda, Marina, Matheus, Nara, Viviane, Pollyanne, Thaís and Tássia. Thank you for teaching me the meaning of the word team! For all your support, from collaborating on data collection and processing to your brilliant literary tips. When the going got tough... you wiped away my tears, held my hand and helped me move on... I'm sure our story doesn't end here, after all, loyal friends like all of you I'll carry in my heart forever! We'll still have great times together!

To my parents, for loving me so much! For being so affectionate with me! Thank you for always making your daughters a priority, for giving up your personal achievements and making my victories your victories! Thank you for giving me a home, moral values and all the supplies so that one day I would be able to walk on my own two feet!

To my sisters Tássia and Thaís, for taking care of me since I was a little girl, always so loving! We lived the best moments of my childhood together and with you I learned to be a less selfish person and to cultivate family values. Thank you for making me believe that time only matures true love!

To my grandparents, Maria Helena and Dagobert Augusto, for teaching me about family and for the love they have shown me since I was born! It was by following your example that I learned more about perseverance and fighting for my dreams;

To my love, Helberte Fernandes Freitas, for accepting me exactly as I am, for encouraging my career, and without any personal pressure accepting my choices that sacrificed some of our moments... You are a wonderful man! Thank you for choosing me to be your partner in life!

To my college friends Érika, Juliana, Leticia, Pat, Fabiane, Loanny, Queiliene, Déborah and Layz, we've had some incredible times together over the five years we've been together and even though we've each gone our separate ways, we've still stuck together! And my childhood friends Zilda and Marcela, always so dear! Many years of support and friendship!

The Coordination, faculty and staff of the Postgraduate Program in Tropical Medicine and Public Health of the Institute of Tropical Pathology and Public Health/UFG, for their dedication, commitment and patience in all stages of this work;

The UFG Social Incubator, especially the coordinator Fernando A. F. Bartholo, for trusting in the work and ethics of our study group, for his willingness to introduce us to each of the recyclable materials cooperatives in Goiania-GO and to support us at all times;

To the staff and teachers at the Faculty of Nursing who welcomed me and taught me the noblest values about nursing, teaching and research;

The board of the qualifying exam, composed of professors Dr. Sheila Araújo Teles, Dr. Márcia Alves Dias de Matos and Dr. Menira Borges de Lima Dias e Souza, for agreeing to contribute scientifically to this study;

To the National Council for Scientific and Technological Development (CNPq) and the Goiás State Research Foundation (FAPEG) for their financial support.

SUMMARY

The hepatitis B virus (HBV) remains one of the main causes of liver disease worldwide, despite the vaccination programs implemented over the last decade. It is estimated that 2 billion individuals have been exposed to HBV and more than 240 million are chronically infected worldwide. Individuals with chronic hepatitis B are at high risk of developing liver cirrhosis and hepatocellular carcinoma. Waste pickers live in precarious social and environmental conditions. Currently, there is little data on HBV infection in this population. Therefore, the general aim of this study was to investigate the epidemiological profile of HBV infection in a population of waste pickers in Goiania, GO. A cross-sectional study was carried out with 431 individuals recruited from 15 recycling cooperatives in Goiania-GO. All participants were interviewed and their serum samples tested for HBV serological markers. Positive HBsAg and anti-HBc samples were subjected to HBV-DNA detection by polymerase chain reaction and genotyped by S-region sequencing. The overall prevalence of HBV infection was 12.8% (95% CI: 9.8-16.2). Multivariate analysis of risk factors showed that age over 40 and illicit drug use were independently associated with HBV infection. Viral DNA was detected in 2/3 HBsAg-positive samples and in 1/52 anti-HBc-reactive samples, resulting in an occult HBV infection rate of 1.9%. Genotypes A (subgenotype A1), D (subgenotype D3) and F (subgenotype F2) were identified. Only 12.3% of the population showed serological evidence of previous vaccination against hepatitis B. These findings highlight the need for public health programs aimed at waste pickers, including vaccination against hepatitis B.

Chapter 1

1. IVI'RODI'CÁO

1.1 A brief history

Viral hepatitis is an infection of the liver caused by five hepatotropic agents: hepatitis A virus (HAV) (Feinstone et al. 1973), hepatitis B virus (HBV) (Dane et al. 1970), hepatitis C virus (HCV) (Choo et al. 1989), hepatitis D virus (HDV) (Rizzetto et al. 1977) and hepatitis E virus (HEV) (Balayan et al. 1983). The sign that indicates hepatitis is jaundice, a yellow-orange coloration of the skin, conjunctivae and mucous membranes caused by high levels of bilirubin in the serum (Hollinger 1996).

Epidemic events involving jaundice were reported by Hippocrates before the Christian era, and evidenced at the end of the 19th century (Hollinger 1996, Hollinger & Liang 2001, Fonseca 2010). In 1885, Lurman described the parenteral transmission of a possible causative agent of hepatitis in shipyard workers, after vaccination against variola prepared with human "lymph" in the city of Bremen in Germany (Lurman 1885 apud Hollinger 1996).

From the 20th century onwards, outbreaks of hepatitis were associated with the use of injectable drugs and blood collection (MacCallum 1972, Fonseca 2010). In 1942, parenteral transmission of a possible agent was suggested, responsible for an epidemic form of hepatitis that affected 28,585 American military personnel, culminating in 62 deaths. This outbreak occurred after the application of a specific batch of yellow fever vaccine stabilized with human serum (Krugman 1989, Reuben 2002). Subsequently, a high frequency of acute hepatitis was observed in individuals transfused with fresh, dried human plasma or whole blood (Morgan & Williamson 1943).

In 1947, MacCallum coined the term "hepatitis B" for an infection with a long incubation period and transmission by blood products and other body fluids (MacCallum et al. 1947). In 1967, Krugman et al. confirmed the existence of the agent proposed by MacCallum, and called it homologous serum hepatitis (MS2) (Krugman et al. 1967 apud Fonseca 2010). The term "hepatitis B" was approved by the Viral Hepatitis Committee of the World Health Organization (WHO 1977) and is still used today.

Blumberg et al. (1965) identified, in a serum sample from an Australian Aborigine, an antigen that reacted specifically with an antibody present in the serum of a hemophiliac patient, calling it "Australia antigen" - AU. In 1968, it was observed that AU (later called hepatitis B virus surface antigen or HBsAg) was found exclusively in the serum of patients with hepatitis B (Prince 1968, Okochi & Murakami 1968, Hollinger 1996). Its identification made it possible to significantly advance research into this hepatitis (Bayer et al. 1968, Almeida et al. 1969).

In 1970, the complete HBV viral particle was identified by electron microscopy, initially called the Dane particle (Dane et al. 1970).

1.2 Hepatitis B virus

1.2.1 Classification and structure of the hepatitis B virus

The hepatitis B virus belongs to the *Hepadnaviridae* family and the *Orthohepadnavirus* genus (ICTV 2011). The complete viral particle or virion is 42 nm in diameter (Figure 1). Externally, it has an envelope containing the L (large), M (middle) and S (small) proteins constituting HBsAg. Internally, the particle is formed by an icosahedral nucleocapsid of approximately 30 nm in diameter, composed of the *core* protein, or *core* antigen (HBcAg), which surrounds the viral DNA and the DNA polymerase/reverse transcriptase enzyme (Hollinger 1996, Bruss 2007, Liang 2009).

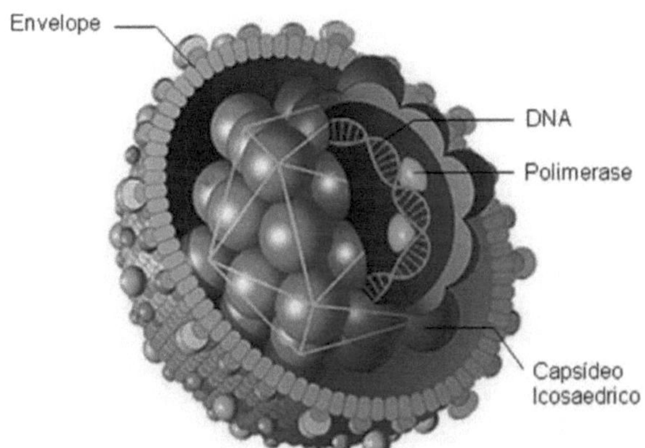

Figure 1- Schematic representation of the structure of the hepatitis B virus Source: http://people.rit.edu/japfaa/infectious.html (modified)

Incomplete, spherical and tubular subviral particles are also secreted by hepatocytes. These lipoprotein particles are approximately 22 nm in diameter (Figure 2) and are identified in the blood of HBV-infected individuals in a concentration 10,000 times greater than virions (Simon et al. 1998, Gilbert et al. 2005, Liang 2009).

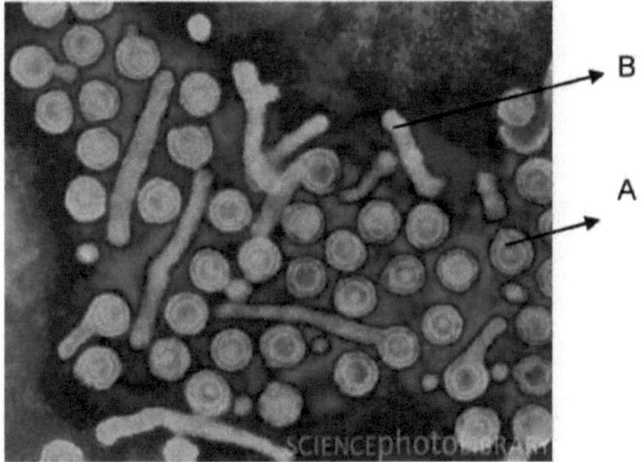

Figure 2- Electron micrograph showing the complete particle (A) and subparticles (B) of HBV Source: http://www.sciencephoto.com/drlindastannard,uct/sciencephotolibrary

HBV has an extremely compact genome made up of partially double-stranded circular DNA with approximately 3,200 base pairs (bp). The smaller strand, known as incomplete, has positive polarity and a 5' to 3' reading direction. The longer strand, called complete, has negative polarity or L (-) and is complementary to the pre-genomic RNA. The circular shape of the HBV DNA molecule is maintained by the pairing of bases at the 5' ends of both strands (Liang 2009, Doo & Ghany 2010). HBV has four *Open Reading Frame (*ORF) sequences in its genome: the Pre-S/S, *Pre-core/core*, P and X regions (Figure 3), which code for viral proteins, and four promoters, which regulate gene activity (Ngui et al. 1999, Seeger & Mason 2000, Hatzakis et al. 2006).

The first ORF is Pre-S/S, which codes for surface proteins called L, M and S, which are synthesized from the start codons of the Pre-S1, Pre-S2 and S regions, respectively, and have the same termination codon at the end of the S region. These proteins play an important role in inducing immunity against HBV and are immunogenic components of the HBV surface (Heermann et al. 1987,

Hourvitz et al. 1996, Le Seyec et al. 1998, Hollinger 2007).

The *Pre-core/core* region, the second ORF, has two start codons responsible for coding the *core* protein (HBcAg) and the "e" protein (HBeAg), both of which induce the production of specific antibodies (anti-HBc and anti-HBe, respectively). The *core* protein, a 21 Kda phosphoprotein, is responsible for assembling the nucleocapsid and is important in the pre-genomic RNA packaging stage. HBeAg is a soluble peptide encoded by the *pre-core/core* regions, which is processed and secreted by liver cells and is considered an important marker of viral replication. This antigen is found during acute infection or in chronic carriers and can induce immune tolerance and, consequently, the development of chronic hepatitis (Milich et al. 1990, Baumert et al. 1996, Hatzakis et al. 2006, Kay & Zoulim 2007, Liang 2009).

The third ORF, region X, encodes the multifunctional HBxAg protein, which acts as a transcriptional activator and is capable of modifying the expression of host cell genes and preventing DNA repair, which would explain the increase in important cell mutations. In addition, it can interfere with the activity of the p53 protein, which has a tumor suppressor and apoptosis activator function and can thus lead to cirrhosis or hepatocellular carcinoma (HCC) in chronically infected patients (Henkler et al. 1995, Kramvis & Kew 1998, Zhang et al. 2001, Bouchard et al. 2004, Wei et al. 2010).

The last ORF, the P or Pol region, is the largest (approximately 75% of the length of the viral genome) and overlaps with the others, coding for the viral polymerase. This has four domains: (1) amino-terminal, which acts as a terminal protein or primase, necessary for starting the synthesis of the negative polarity DNA strand; (2) spacer, which has no well-defined function; (3) reverse transcriptase, responsible for transcribing pre-genomic RNA into DNA and (4) carboxy-terminal, which exhibits ribonuclease H (RNAse H) activity, participating in the degradation of pre-genomic RNA (Seeger & Mason 2000, Block et al. 2007, Liang 2009).

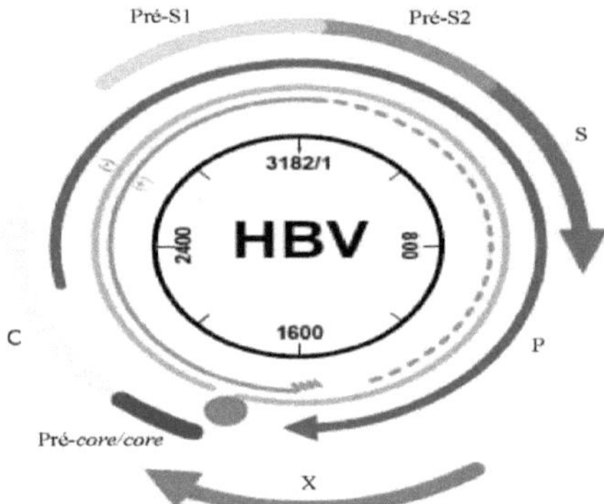

Figure 3- Schematic representation of the HBV genome with the ORFs: Pre-S/S, *Pre-core/core*, P and X

Source: Neuveut et al. (2010) (modified)

1.2.2 HBV variability

Serological subtypes

Studies carried out in the early 1970s identified antigenic variations in HBsAg, defined by two pairs of mutually exclusive allelic determinants, *d/y* and *w/r*, which together with the "*a*" determinant, resulted in four different serological subtypes of HBV: *adw*, *ayw*, *adr* and *ayr* (Le Bouvier 1971, Bancroft et al. 1972). Furthermore, the identification of the *w* subdeterminant (w_{1-w4}) and the *q* determinant (q+/q-) allowed for a broader classification into nine serological subtypes: adw_2 , adw_4, $aywl$, ayw , ayw_{23} , ayw_4 , $adrq+$, $adrq-$ and *ayr* (Figure 4) (Courouce et al. 1976, Courouce-Pauty et al. 1978, 1983).

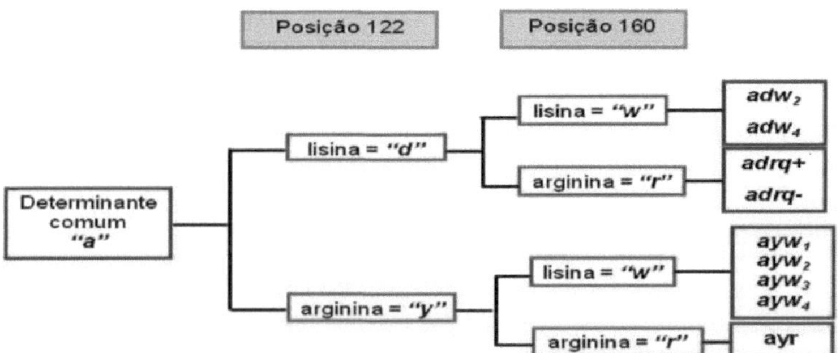

Figure 4- HBV serological subtypes
Source: Adapted from Kidd-Ljunggren et al. (1994)

Genotypes

The divergence of more than 8% in the complete HBV genome sequence allowed the identification of eight genotypes, classified from A to H (Figure 5) (Okamoto et al. 1988, Norder et al. 1992, 1993, 1994, Naumann et al. 1993, Stuyver et al. 2000, Arauz-Ruz et al. 2002, Kubarnov et al. 2010). Isolates of HBV have been identified in Vietnam that differ from the known genotypes, and it has been suggested that they be called genotype I (Huy et al. 2008, Tran et al. 2008, Yu et al. 2010). Subsequently, the "J genotype" was identified by Tatematsu et al. (2009) in patients from Japan. However, these new HBV genotypes have been questioned and further studies are needed to definitively classify them (Kubarnov et al. 2010).

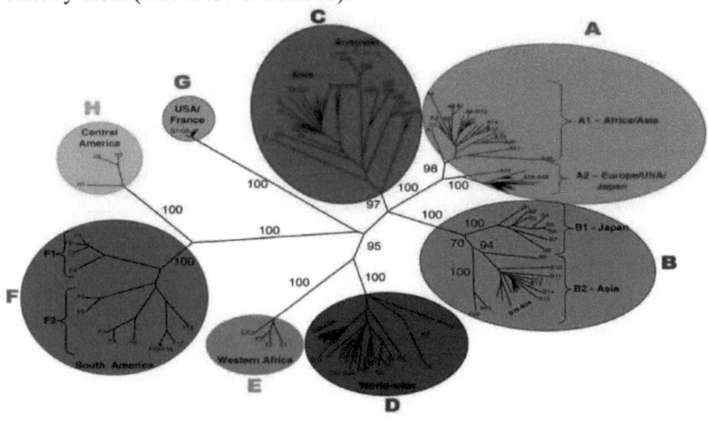

Figure 5- Phylogenetic tree with the eight HBV genotypes (A-H) Source: Kramvis et al. (2005) (modified)

The characterization of genotypes is relevant in relation to the clinical aspects of the disease, as they influence antiviral therapy and the progression of liver disease. They are also of epidemiological importance, elucidating the dynamics of HBV infection and transmission (Chu & Lok 2002, Kao 2002, Schaefer 2005).

Subgenotypes

Genotypes can also be divided into subgroups with differences of between 4% and 8% within the same genotypic group, known as subgenotypes, and differences of less than 4% are known as *clades*. Subgenotypes are designated by alphabetical letters followed by corresponding numbers (Kramvis et al. 2005).

Thus, some genotypes are divided into subgenotypes, such as: A (A1 - A7), B (B1 - B9), C

(C1 - C12), D (D1 - D8) and F (F1 - F4) (Norder et al. 2004, Huy et al. 2006, Kramvis et al. 2008, Mulyanto et al. 2009, 2011, Abdou Chekaraou et al. 2010, Kurbanov et al. 2010, Hübschen et al. 2011, Thedja et al. 2011).

Further studies are necessary since some of the subgenotypes identified are still considered provisional, which show a small nucleotide divergence compared to others and can be considered *clades of the* same subgenotype (Kurbanov et al. 2010).

1.2.3 Viral replication

The HBV replicative cycle is shown in Figure 6. Adsorption and penetration of the viral particle occur in the host's hepatocytes; however, these mechanisms are not fully understood. It is believed that the L protein has a recognition site for adsorption of the virus to a receptor located on the surface of the target cell (Hollinger 2007). After virus-cell interaction, penetration occurs by endocytosis. The virus loses its envelope, releasing the nucleocapsid containing DNA, which passes through the nuclear pores into the cell nucleus (Kann et al. 1997, Beck & Nassal 2007).

HBV-DNA, already inside the nucleus, is converted into *covalently closed* circular double-stranded DNA (cccDNA) by the action of DNA polymerase (Beck & Nassal 2007), which then serves as a template for the transcription of pgRNA (3.5 Kb), essential for replication, and other viral mRNA molecules (0.7, 2.1 and 2.4 Kb). This transcription occurs through the action of the cellular enzyme RNA polymerase II (Gonçales Jr 2002, Ganem & Prince 2004, Beck & Nassal 2007, Levrero et al. 2009, Liang 2009).

The mRNA strands are transported to the cytoplasm and translated into viral surface proteins, *core,* polymerase and X. In the cytoplasm, the nucleocapsid is formed by the assembly of pgRNA together with DNA polymerase/reverse transcriptase. Transcription of the pgRNA inside the viral nucleocapsid then begins, forming the first long-chain (negative) strand of viral DNA. During or after the synthesis of this strand, the pgRNA is degraded by the enzymatic action of RNAse H polymerase and, by the action of DNA polymerase, the synthesis of the positive or incomplete strand of viral DNA occurs (Gongales Jr 2002, Ganem & Prince 2004, Beck & Nassal 2007, Hollinger 2007, Levrero et al. 2009).

Subsequently, the nucleocapsid acquires the envelope in the rough endoplasmic reticulum. The virions are then released from the hepatocytes. However, some DNA molecules are transported back to the nucleus and can be converted into cccDNA, resulting in intracellular viral persistence (Seeger & Mason 2000, Ganem & Prince 2004, Beck & Nassal 2007). According to Hollinger et al. (2007) approximately 10^{11} molecules/day are produced during viral replication, and during peak replication this rate can increase 100 to 1000 times.

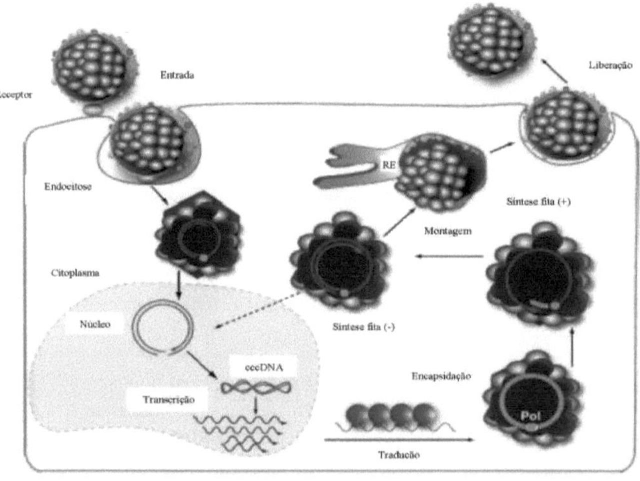

Figure 6 - Diagram of the HBV replicative cycle Source: Neuveut (2010) (modified)

1.3 Clinical Aspects, Pathogenesis and Treatment of HBV Infection

Hepatitis B is one of the most common and serious infectious diseases due to its high morbidity and mortality, as the infection can cause acute and chronic liver disease, including liver cirrhosis and hepatocellular carcinoma (HCC) (Kao & Chen 2002, Chang 2007, Chang & Lewin 2007, Liang 2009). Around two billion individuals have been exposed, and approximately 240 million are chronic HBV carriers worldwide. It is estimated that 600,000 individuals die each year from the acute or chronic consequences of hepatitis B (WHO 2012a).

Acute HBV infection can be asymptomatic or symptomatic. The symptomatic form comes in two forms: benign (long course of the disease, with complete recovery from liver damage) and severe or fulminant. According to the evolution of the disease, acute hepatitis can be classified into four phases: the incubation period, the prodromal or pre-icteric phase, the icteric phase and the convalescent phase (Focaccia 2002, Chang 2007).

The incubation period ranges from 45 to 180 days. The prodromal phase is characterized by non-specific symptoms, similar to those of the flu, such as low-grade fever, fatigue, anorexia, myalgia, nausea and vomiting, abdominal pain, among others (Hollinger 1996, Befeler & Di Bisceglie 2000, Gonçales Jr 2002, Chang 2007).

In the icteric phase, choluria appears due to the increase in bilirubin levels in the blood, causing fecal acolyria and yellowing of the mucous membranes of the conjunctiva (sclera) and skin, a situation clinically recognized in 20% of infected patients. Serum levels of bilirubin (mainly the direct fraction) and aminotransferases (aspartate aminotrasferase - AST and alanine aminotrasferase - ALT) are elevated, the latter being associated with the presence of liver damage. This phase usually lasts around 20 days and manifests itself 10 days after the onset of symptoms in approximately 85% of cases (Molner & Meyer 1940, Zuckerman 1965, Mendonça & Vigani 2006, Chang 2007).

The convalescence phase lasts an average of 20 to 30 days, with progressive improvement in clinical symptoms, with a slow reduction in hepatosplenomegaly, jaundice and dyspeptic symptoms, as well as viral elimination (Mendonça & Vigani 2006).

Fulminant hepatitis is characterized by rapid progression to liver failure, the presence of hepatic encephalopathy, jaundice, coagulopathy and high levels of aminotransferases. It has a high mortality rate, occurring in 1% of individuals with acute hepatitis B (Inoue et al. 1998, Petrosillo et al. 2000, Carey 2009).

Progression to chronic HBV infection depends on factors such as the age at which the infection was acquired, ongoing HBV replication and the patient's immune response. Chronic HBV infection affects 5-10% of adults exposed to HBV in adulthood. Among children infected in the first year of life, 90% become chronic carriers and, among those infected between one and four years of age, 30-50% develop chronic hepatitis B (De Franchis et al. 1993, Maruyama et al. 1993, Maddrey 2001, Jung & Pape 2002, Chang 2007, WHO 2012a).

Some factors can influence the progression of liver disease in individuals with hepatitis B, including gender (male), alcohol consumption, family history of HCC, childhood HBV infection, seropositivity for HBeAg and viral genome (cccDNA), as well as the presence of co-infection with other hepatotropic viruses (Baffis et al. 1999, Mendonga & Vigani 2006, Nwokediuko 2011, El-Serag 2012). In addition, chronic infection can increase the chance of developing cirrhosis and hepatocellular carcinoma (Kew 1998, Arbuthnot & Kew 2001, Feitelson & Lee 2007, Negro 2011).

HBV infection is responsible for inducing tissue damage of varying severity, without generating a direct cytopathic effect on liver cells (Ferrari et al. 2003, Chang & Lewin 2007). Damage to liver tissue begins with stimulation of the host's cellular and humoral immune response against specific viral antigens present on the surface of cells infected by the virus (Ferrari et al. 2003, Mello & Alves 2006).

In terms of cellular immunity, infected cells are lysed by CD8 cytotoxic T lymphocytes[+], which are responsible for the response directed against multiple epitopes of the HBV *core*, polymerase and envelope proteins (Rehermann et al. 1996, Rapicetta et al. 2002, Ganem & Prince 2004, Vierling 2007). However, cell destruction is not the only strategy capable of eliminating HBV

(Ferrari et al. 2003). Through non-cytolytic mechanisms, cytotoxic T lymphocytes release cytokines such as interferon-gamma (IFN-y) and tumor necrosis factor-alpha (TNF-a) in the liver, which act in antiviral protection, regulating HBV replication without destroying the infected cell. In this way, IFN-y inhibits the gene expression of the virus and is mainly responsible for the recruitment and activation of innate immune response cells in the liver, such as macrophages and *natural killer* cells. Another mechanism involves the humoral immune response, where B lymphocytes produce specific antibodies, responsible for forming a complex with the free viral particles, preventing infection of susceptible hepatocytes (Rapicetta et al. 2002, Ferrari et al. 2003, Ganem & Prince 2004, Inchauspé & Michel 2007).

Occult infection, a peculiar form of chronic infection, is mainly characterized by the persistence of HBV-DNA in HBsAg-negative individuals. This manifestation of hepatitis B is related to the persistence of cccDNA in the liver and a strong suppression of viral replication and gene expression (Conjeevaram & Lok 2001, Raimondo et al. 2007, Hollinger & Sood 2010, Larrubia et al. 2011, Raimondo 2012).

The mechanism of occult HBV infection has not been fully elucidated. Some hypotheses have been suggested, such as mutations in the S, *core* and X regions, integration of the viral genome with that of the host, formation of immune complexes, alteration of the immune response pattern, superinfection and mixed infection with interference from HBV variants (Ferreira 2000, Lok 2004, Fonseca 2007, Lledó et al. 2011, Romero et al. 2011, Raimondo et al. 2012).

The impact of occult infection can be seen in different clinical contexts. It can be transmitted by blood transfusions and liver transplants, causing classic forms of hepatitis B in infected individuals. In immunosuppressed patients (mainly due to immunotherapy or chemotherapy), there may be reactivation and development of acute hepatitis with a more severe clinical course. Evidence suggests progression to cirrhosis, as well as hepatocellular carcinoma. In those co-infected with HCV, some authors have also shown progression to fibrosis and changes in liver enzymes (Liu & Kao 2007, Matsuoka et al. 2008, Selim et al. 2011, Fuente et al. 2011, Wong et al. 2011, Shi et al. 2012).

Treatment of HBV infection aims to reduce the risk of liver disease progression and its primary outcomes, specifically cirrhosis, hepatocarcinoma and, consequently, death. Surrogate or intermediate outcomes, such as the level of HBV-DNA, liver enzymes and serological markers, are validated and have been used as parameters to infer the likelihood of long-term benefits from therapy, given that sustained suppression of viral replication and a reduction in histological activity reduce the risk of cirrhosis and hepatocarcinoma (Papatheodoidis et al. 2002, Coffin & Lee 2009, Marcelim et al. 2009, BRASIL 2011b).

Individuals with acute infection do not need specific antiviral treatment, as around 90% of adults are spontaneously cured within a period of approximately three months. For patients with fulminant hepatitis B, liver transplantation is recommended (EASL Jury 2003, Chang & Lewin 2007, Pérez 2007, Beckebaum et al. 2009, Shiffman 2010).

There are two categories of antivirals used in the treatment of hepatitis B: immunoregulatory agents (conventional interferon alpha and pegylated interferon alpha) and nucleoside analogues (lamivudine, entecavir and telbivudine) or nucleotide analogues (adefovir and tenofovir) which act by inhibiting the reverse transcription of HBV (Xu & Chen 2006, Chang & Lewin 2007, Pawlotsky et al. 2008, Coffin & Lee 2009).

In Brazil, the Ministry of Health's therapeutic protocol for chronic HBV infection indicates the following drugs: interferon-alpha, pegylated interferon-alpha, lamivudine, tenofovir, entecavir and adefovir, in the following clinical situations: 1) Treatment-naive, HBeAg reactive, non-cirrhotic individuals, 2) Treatment-naive, HBeAg non-reactive, non-cirrhotic individuals, 3) Treatment-naive, cirrhotic individuals, HBeAg reactive or non-reactive, 4) Patients experienced with antivirals (resistance to antivirals) and 5) Special situations (chronic viral hepatitis B in children and HBV-HDV, HBV-HIV and HBV-HCV co-infections (BRASIL 2011b).

1.4 Laboratory Diagnosis of HBV Infection

HBV infection can be diagnosed by serological tests capable of detecting the presence of specific markers, such as antigens (HBsAg and HBeAg) and antibodies (anti-HBs, anti-HBe and anti-

HBc), as well as by testing for viral DNA using molecular biology techniques (Gonçales & Cavalheiro 2006, Gonzales & Salinas 2009). Liver assessment can also be carried out by measuring bilirubin, alkaline phosphatase, gamma-glutamyltransferase (y-GT), AST and ALT, which indicate hepatocellular damage (Fonseca 2007, Liang 2009, Liaw & Chu 2009). The progression of the disease can also be monitored through histological examination of the liver biopsy, assessing inflammation and necrosis, as well as damage related to fibrosis and hepatocellular carcinoma (Yun-Fan & Chia Ming 2009).

HBsAg, the first serum marker detected in HBV infection, can be present before the onset of symptoms and usually appears around the fourth week after exposure to the virus (Figure 7). In patients who recover from the infection, HBsAg titers decline and disappear within five to six months. After this period, specific antibodies to this antigen (anti-HBs) develop, the presence of which indicates clinical recovery and immunity to HBV (Hollinger 1996, Ferreira 2000, Badur & Akgün 2001, Khouri & Santos 2004, Shiffman 2010).

HBcAg is only detected in infected hepatocytes and is therefore not found in patients' serum. However, antibodies directed against it (anti-HBc) are routinely detected. The anti-HBc IgM marker is usually detectable at the onset of symptoms, appearing shortly after HBsAg and disappearing as the infection resolves. Total anti-HBc (IgG), on the other hand, persists throughout life and is considered a marker of exposure to HBV (Ferreira 2000, Badur & Akgün 2001, Gonçales & Cavalheiro 2006, Hatzakis et al. 2006, Chevaliez & Pawlotsky 2008, Shiffman 2010).

Another antigen that appears at the beginning of the acute phase is HBeAg, the presence of which is associated with viral replication and infectivity. However, viral replication can occur in its absence, as mutations in the *pre-core* region or in the *core* promoter prevent the synthesis of this antigen. In response to HBeAg, anti-HBe antibodies are detected, indicating a decrease in viral replication. During acute infection, these antibodies are associated with a probable spontaneous resolution of the infection (Ferreira 2000, Badur & Akgün 2001, EASL Jury 2003, Gonçales & Cavalheiro 2006, Chevaliez & Pawlotsky 2008, Shiffman 2010).

The presence of viral DNA in the patient's serum occurs within a few days of the onset of infection and is considered the most reliable marker of present infection, as well as being an important parameter for therapeutic decisions and monitoring the response to treatment. During acute infection, HBV-DNA levels increase, reaching a peak, followed by a progressive decline in this marker, the disappearance of which indicates spontaneously resolved acute infection. Sensitive molecular methods can detect HBV-DNA 10 to 20 days before HBsAg is detected (EASL Jury 2003, Pawlotsky 2003, Diaz & Mendonça 2006, Gonçales & Cavalheiro 2006, Shiffman 2010).

Molecular techniques that detect viral DNA, such as polymerase chain reaction (PCR), are widely used to diagnose HBV infection. In addition, quantitative research (detection of viral load), using the real-time PCR technique, has been used to verify the evolution of the infection (Van Deursen et al. 1998, Ferreira 2000, Pawlotsky 2003, Valsamkis 2007, Seto et al. 2011).

In addition, HBV genotyping methods such as PCR amplification with genotype-specific *primers*, RFLP (*Restriction* Fragment *Length Polymorphism*), hybridization after PCR (LiPA) and serotyping, as well as viral genome sequencing have been used (Valsamakis 2007, Seto et al. 2011).

Chronic hepatitis B infection can be diagnosed by the persistence of HBsAg for six months or more, and the markers anti-HBc total and HBeAg or anti-HBe can also be detected (Figure 8) (Ferreira 2000, Liang 2009, Liaw & Chu 2009).

Different serological patterns and serum levels of HBV-DNA are related to the stages of development of chronic infection (EASL Jury 2003, Pawlotsky 2003, Liaw & Chu 2009). The first is defined as the immunotolerance phase. This phase commonly occurs after perinatal transmission and is characterized by the presence of serum HBsAg (for a period of more than 6 months), HBeAg, high titres of HBV-DNA (10^{5-10} copies per mL), ALT at normal or slightly elevated levels, minimal histological liver damage and an asymptomatic course. It is suggested that the primary function of HBeAg is to induce a state of immunotolerance in HBV carriers (HBsAg reactive). In individuals exposed to HBV in childhood, the immunotolerance phase can last for one to four decades. However, when adults become infected with HBV, this phase is not observed. Patients with this phase are considered to be at low risk of progression to liver cirrhosis and hepatocarcinoma (Lok & McMahon 2001, Raimondo et al. 2003, Mendonça & Vigani 2006, Fonseca 2007, Elgouhari et al. 2008, Liaw & Chu 2009).

The second phase, called immunoactive, is characterized by the presence of HBeAg (wild HBV) or anti-HBe (residual wild HBV or *pre-core* mutant) in the serum. This phase occurs after horizontal transmission between children or in adulthood, and also late in life among people who have acquired HBV infection through vertical transmission, starting right after the immunotolerance phase. High levels of ALT and HBV-DNA and active liver disease seen on biopsy characterize this phase. Patients with HBeAg-positive chronic hepatitis B may present spontaneous seroconversion to anti-HBe, with a rise in ALT. After seroconversion, normal ALT levels and HBV-DNA titers of less than 1000 IU/mL (10^3 copies/mL) are observed (Lok & McMahon 2001, Raimondo et al. 2003, Mendonga & Vigani 2006, Fonseca 2007, EASL 2009, Liaw & Chu 2009, McMahon 2010).

In the third phase, known as inactive HBV carrier, HBsAg, anti-HBe, low or undetectable HBV-DNA titers, normal ALT, minimal histological liver damage, an asymptomatic course and a good prognosis. Many inactive HBV carriers (70% to 90%) remain inactive for life. An additional 10% to 20% of inactive carriers show reversion phenomena, characterized by the reappearance of HBeAg, usually accompanied by an increase in ALT due to the process of inflammatory reactivation of the liver. A much smaller number of inactive HBV carriers develop anti-HBe positive chronic hepatitis B (residual chronic hepatitis due to wild HBV), which is characterized by elevated aminotransferase levels, HBV-DNA titer greater than 20000 IU/L (>105 copies m/l) and active (histological) liver disease. However, the clinical course and sequelae of chronic hepatitis caused by wild or mutant HBV vary from individual to individual (Lok & McMahon 2001, Fattovich 2003, Gongales & Gongales Jr 2006, Mendonga & Vigani 2006, Fonseca 2007, Liaw & Chu 2009).

Reactivation of infection can be considered the fourth phase of HBV infection, characterized by the reappearance of necroinflammatory activity in the liver in inactive HBV carriers or in those diagnosed with resolved infection (previous HBV infection, without virological or biochemical signs or histological evidence of active viral disease) (Lok & McMahon 2001, Fonseca 2007, EASL 2009, Liaw & Chu 2009).

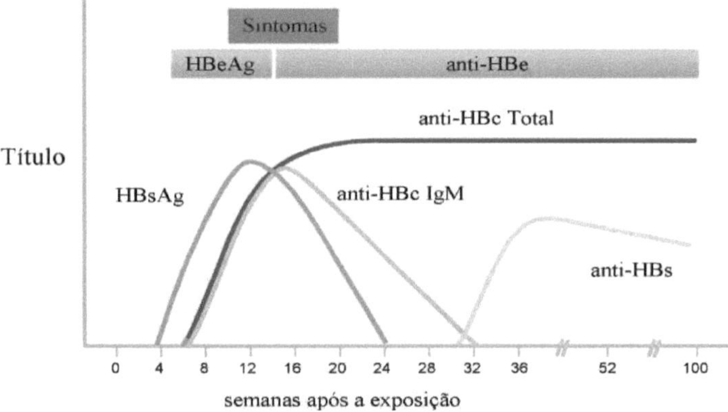

Figure 7- Serological profile of acute hepatitis B
Source: CDC - http://www.cdc.gov/ncidod/diseases/hepatitis/slideset (modified)

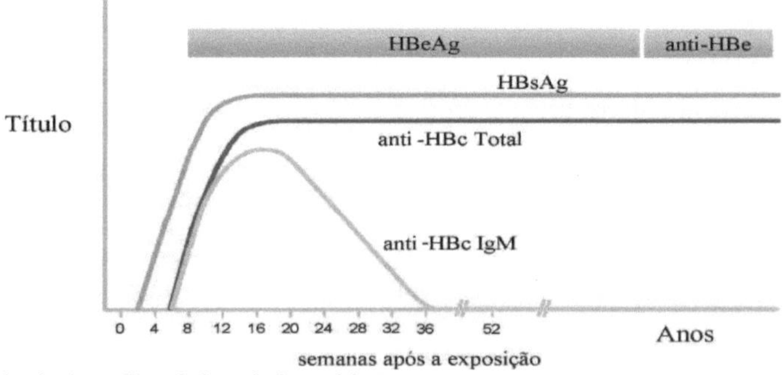

Figure 8- Serological profile of chronic hepatitis B

Source: CDC - http://www.cdc.gov/ncidod/diseases/hepatitis/slideset (modified)

1.5 Epidemiology of HBV Infection

1.5.1. Hepatitis B virus transmission

HBV transmission routes include vertical, sexual and parenteral/percutaneous (Hou et al. 2005, Kwon & Lee 2011). Although HBsAg has been found in various body fluids, only serum, semen and saliva have been shown to be infectious (Alter 2003). HBV is stable on environmental surfaces for more than seven days (Bond et al. 1981), and consequently indirect transmission of the virus by contaminated objects can occur. The risk of transmission is higher when the level of HBV-DNA is high in the serum, especially in HBeAg reactive patients (Alter et al. 1976, Alter et al. 2003, Elgouhari et al. 2008, Shiffman 2010, Kwon & Lee 2011).

The vertical route is characterized by the transmission of HBV during pregnancy (transplacental), at the time of delivery (contact with contaminated blood/amniotic fluid) and afterwards (Jonas et al. 2009).

On the other hand, the spread of HBV by the sexual route makes it possible to classify hepatitis B as a sexually transmitted disease (STD). This mode of transmission occurs mainly in low-prevalence regions, among adolescents and adults with risk behaviors, such as multiple sexual partners, non-use or occasional use of condoms and a history of other STDs (Alter 2003, Hou et al. 2005, Carey 2009, Lee & Park 2010, Shiffman 2010, Franco et al. 2012).

The parenteral/percutaneous route has blood or blood products containing HBV as a source of contamination. People at risk of infection via this route are patients who need blood transfusions (hemodialysis patients, hemophiliacs and people with neoplasms), as well as health professionals and users of illicit drugs, especially injectable drugs. In addition, *piercing*, tattooing and acupuncture with contaminated materials are risk activities for acquiring HBV (Teles et al. 2002, Ferreira et al. 2009, Shiffman 2010, Niederhauser 2011, WHO 2012b). This infection can also occur through sharing sharp objects for personal use (razors, toothbrushes, manicure pliers, etc.) (Hou et al. 2005, Lavanchy 2008, Elgouhari et al. 2009, Shiffman 2010, Mu et al. 2011).

1.5.2. Distribution of HBV infection

Ott et al. (2012) published a systematic review of the literature on the global epidemiology of HBV infection, based on 396 HBsAg prevalence studies. To illustrate these data, two maps on the prevalence of HBV infection in children aged 5-9 (Figure 9) and adults aged 19-45 (Figure 10) are presented.

In areas of high endemicity for HBV infection (HBsAg prevalence greater than or equal to 8%), vertical transmission occurs mainly in the perinatal period or during childhood. These regions are considered underdeveloped and have a high population density, such as sub-Saharan Africa, the western part of which had the highest age-specific prevalence in the world in 1990, reaching up to 12% HBsAg positivity among children and adolescents up to the age of 19. Although there was a decrease in 2005, the region continues to have high endemicity, especially in males (Ott et al. 2012).

In regions of intermediate endemicity, HBsAg prevalence ranges from 27%. East Asia had the highest HBV prevalence and there were no significant changes between 1990 and 2005. Overall, endemicity remained high-intermediate in this region, mainly in males. In South Asia, around 3% of the population up to the age of 45 was HBsAg positive, with a reduction in prevalence in older individuals (low prevalence in 2005). Unlike other regions, in Southeast Asia, there was a strong reduction in HBsAg prevalence between 1990 and 2005, especially in the 0-14 age group, with a prevalence of 1.21.4%. In high-income countries, including Japan, the Republic of Korea and Singapore, there has also been a significant reduction, and endemicity is considered intermediate, with a prevalence of approximately 4% (Ott et al. 2012).

The Pacific Islands and the Oceania region were endemic in 1990, reaching a rate for HBsAg of around 10% in men aged 10-34. The decrease in prevalence has led to a shift to high intermediate endemicity in the age groups up to 54 years, and intermediate endemicity in older adults (Ott et al. 2012).

HBsAg seroprevalence has been shown to be low in Western Europe, particularly in women, whose prevalence was below 2% between 1990-2005. However, during this period, an increase was observed in both sexes, which caused a change in endemicity to low-intermediate in young men in 2005 and a decrease in prevalence in older individuals. In Central Europe, the prevalence in infants and girls decreased from 6% to 3%. In contrast, Eastern European countries did not show a strong reduction in HBsAg prevalence at younger ages. In Central and Eastern Europe, ages 0-9 remain the most affected by HBsAg infection (Ott et al. 2012).

In high-income countries in North America (Canada and the United States), prevalence declined in both sexes and all ages between 1990 and 2005. The highest positivity for HBsAg was in male children aged between 04 years (2.14%), and the lowest (around 1%) in individuals over 65 in 2005 (Ott et al 2012).

In Latin America, there was a sharp decrease in HBsAg prevalence between 1990 and 2005. In the tropical region, endemicity went from intermediate to low, where boys aged 0-9 had the highest prevalence in the region (intermediate endemicity), over 5% in 1990, and 1.6% in 2005. Similarly, the prevalence in the central region declined by half in this period, and in adults, it was considered to be of low endemicity in 2005. Other Latin American regions, such as the Andes and South America, have shown a decrease in prevalence with age and intermediate endemicity (Ott et al. 2012).

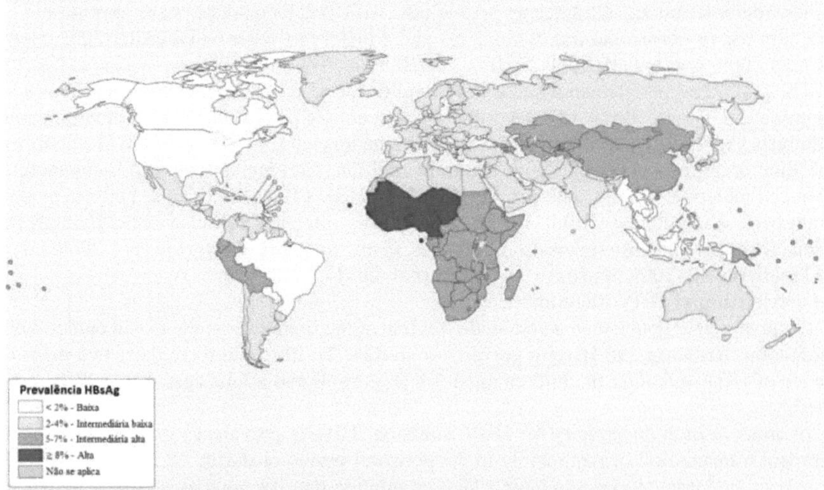

Figure 9 - Prevalence of HBV infection in children aged 5-9, 2005
Source: Ott et al. (2012) (modified)

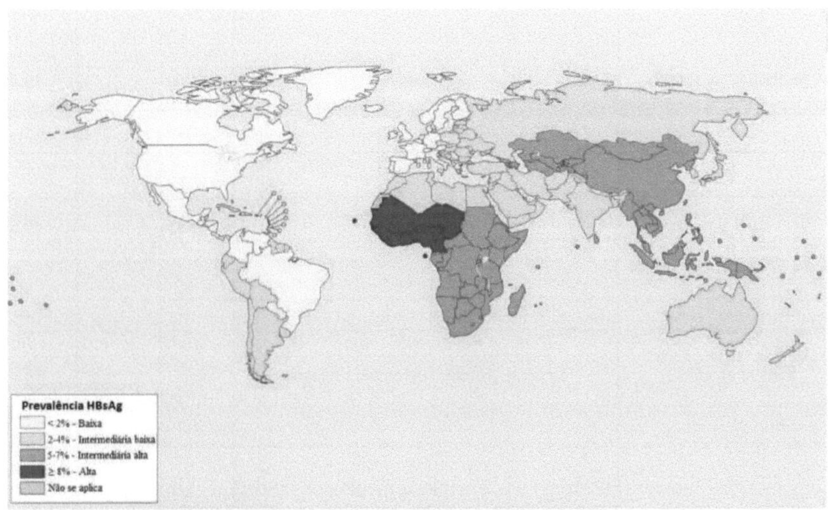

Figure 10 - Prevalence of HBV infection in adults aged 19-49, 2005
Source: Ott et al. (2012) (modified)

In Brazil, a population-based survey carried out in Brazilian state capitals showed a prevalence for the HBsAg marker of 0.37% (95% CI: 0.25-0.50), 0.05% (95% CI: 0.01-0.10) in the 10-19 age group and 0.6% (95% CI: 0.410.78) in the 20-69 age group. The overall prevalence (anti-HBc), also for all of Brazil's capital cities, was 7.4% (95% CI: 6.8-8.0). The percentage of those exposed to HBV in the 10-19 age group was 1.1% (95% CI: 0.9-1.4) and 11.6% (95% CI: 10.7-12.4) for the 20-69 age group. In all regions, there was an increase in total anti-HBc positivity with age. With regard to gender, men were more likely to be exposed in all regions and in the Federal District, except in the North. There was a higher risk of exposure to HBV in individuals with poorer socioeconomic conditions, except in the Southeast. Sexual transmission was relevant in the North, Northeast, Midwest and South, and in the latter, blood transmission also stood out (BRASIL 2011a).

Variable prevalence rates of HBV infection have been observed in the different populations studied in Goiania-GO (Table 1).

Table 1 - Prevalence of HBV infection in studies carried out in the metropolitan region of Goiania-GO

Referendum	Populated	N	Prevalence	
			HBsAg(%)	Overall (%)
Cardoso et al. (1990)	Female population	475	-	6,1

Martelli et al. (1990)	Prisoners	201	-	26,4
	Primodoadores	1033	-	12,8
Rosa et al. (1992)	Leprosy patients undergoing treatment	83	4,8	16,9
	outpatient			
	Institutionalized leprosy patients	171	8,8	50,5
Azevedo et al. (1994)	Health professionals	625	2,3	23,4
Porto et al. (1994)	Street children	496	2	13,5
Cardoso et al. (1996)	Pregnant women	1459	0,5	7,5
Borges et al. (1997)	Dialysis patients	175	13,7	63,4

Teles et al. (1998)	Hemodialysis patients	282	12	56,7
Bastos (2000)	Patients with sickle cell disease	63	0	12,7
Lopes et al. (2001)	Dialysis center professionals	152	0,7	24,3
Silva et al. (2002a)	Individuals with suspected hepatitis	1396	14,5	50,7
Teles et al. (2002)	Hemodialysis patients	536	5,8	-
Carneiro & Daher (2003)	Anesthesiologists	90	0	8,9
Costa et al. (2004)	Elderly residents of nursing homes	195	0,5	32,3
Souza et al. (2004)	People with mental illnesses	433	1,6	22,4
Tavares et al. (2004)	Hemophiliacs	102	1,1	43,7

Silva et al. (2005b)	Laboratory professionals	648	0,7	24,1
Oliveira et al. (2006)	Low-income schoolchildren	644	0,6	5,9
Ferreira (2008)	Illicit drug users	422	0,7	14,7
Paiva et al. (2008)	Dentists	680	0	6,0
Aires et al. (2012)	Tuberculosis patients with or without HIV	402	1,2	25,6

1.5.3. Distribution of genotypes

The HBV genotypes have a geographically distinct distribution. Thus, genotype A is prevalent in the United States, parts of Europe and India, and is also found in the Philippines, eastern, southern and central Africa. Genotypes B and C have a similar geographical distribution, being detected in populations in southwest Asia, Australia, Japan, China and Vietnam. Genotype D is distributed in the Mediterranean, India and western Europe. Genotype E is restricted to Africa F circulates in South and Central America and Polynesia. Genotype G was initially identified in samples from North America and part of Europe and, finally, genotype H circulates in the United States, Mexico and Central America (Figure 11) (Kay & Zoulim 2007, Dehesa-Violante & Nuñez-Nateras 2007, Mahtab et al. 2008, Jazayeri et al. 2010, Kew 2010, Kurbanov et al. 2010, Te & Jensen 2010, Kao 2011).

In Brazil, studies have shown the circulation of genotypes A, B, C, D, F and G. However, genotypes A, D and F predominate (Mello et al. 2007, Ferreira et al. 2009, Matos et al. 2009, Santos et al. 2010, Alvarado-Mora 2011, Pinho et al. 2011, Ramos et al. 2011, Scaraveli et al. 2011, Silva et al. 2011, Aires et al. 2012, Bertolini et al. 2012, Carvalho et al. 2012, Dias et al. 2012, Matos et al. 2012).

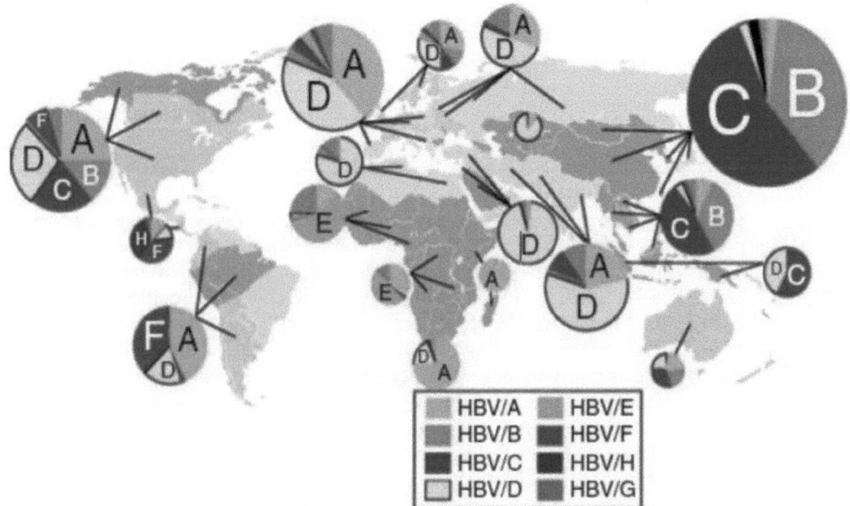

Figure 11 - Geographical distribution of HBV genotypes Source: Kurbanov et al. (2010)

1.6 Prevention and Control of HBV Infection

Hepatitis B virus infection can be prevented through behavioral changes regarding the risk of parenteral transmission, such as not sharing sharp personal hygiene objects (toothbrushes, razors, nail clippers, etc.) and syringes/needles, syringes/needles, as well as screening for HBV in blood banks and making it clear that HBV carriers and those at recent risk of infection cannot donate blood, organs, tissues or semen. Another important measure is the adoption of biosafety practices in health establishments, as well as adequate guidance for employees regarding the use of PPE (Personal Protective Equipment) and post-accident immunoprophylaxis (Ferreira & Silveira 2004, Hou et al. 2005, BRASIL 2006b, Van Herck 2008, Kwon & Lee 2011, Niederhauser 2011).

Serological screening of blood donors (HBsAg and anti-HBc in some countries) is an important prevention strategy. In addition, some countries have included the detection of HBV-DNA in order to further reduce the risk of post-transfusion infection (Niederhauser 2011). Improving the methods used for viral inactivation of blood products is also an important measure to minimize the risk of HBV transmission (Hou et al. 2005, BRASIL 2006a). Serological screening of pregnant women is recommended in Brazil, as part of the prenatal care program as part of the list of so-called "Breast Tests", as well as prophylaxis in newborns of HBsAg reactive mothers (BRASIL 2008, Jonas 2009).

In order to prevent the sexual transmission of HBV, counseling on the use of condoms, avoiding sexual relations with multiple partners, both homosexual and heterosexual, among others, should be emphasized (Van Herck et al. 2008, Liaw & Chu 2009).

The most effective and safe prevention measure for controlling hepatitis B is vaccination (Joshi & Kumar 2001, Lavanchy 2004, BRASIL 2005, 2006b, CDC 2007, Kwon & Lee 2011). Available since 1982, the first generation of hepatitis B vaccine, derived from the plasma of HBV carriers, is composed of inactivated subviral particles (Joshi & Kumar 2001, Shepard et al. 2006, Lee & Park 2010).

With the advance of recombinant DNA technology, the second-generation vaccine was developed, consisting of purified recombinant HBsAg adsorbed to an adjuvant, which became available in 1986 (Joshi & Kumar 2001, Koff 2003, Hou et al. 2005, CDC 2007). Administered intramuscularly, the hepatitis B vaccine should be administered following a three-dose schedule, with a one-month interval between the first and second doses, and the third dose should be administered six months after the first dose (0, 1 and 6 months), with the induction of protective antibodies occurring in over 90% of healthy adults and young people, and over 95% in infants, children and adolescents (Bruguera 2006, Shepard et al. 2006, Van Herck et al. 2008, Alavian et al. 2012).

Vaccine response is assessed by the detection of the anti-HBs marker, which must be present at levels greater than or equal to 10 mUI/mL. However, immunity can persist after the disappearance of these antibodies over the years. This phenomenon is explained by immunological memory, with anti-HBs levels rising after exposure to HBV. For this reason, a booster dose of the vaccine is not recommended for immunocompetent adults (Hou et al. 2005, Shepard et al. 2006, Van Herck et al. 2008, Aspinall et al. 2011, Alavian et al. 2012).

According to the WHO, 177 countries have included the three-dose vaccination schedule against hepatitis B in the vaccination calendar, thus reducing the incidence of HBV in these places (WHO, 2010). In Brazil, this vaccine is part of the National Immunization Program (PNI) and is recommended for newborns on the first day of life, children, adolescents and adults up to 29 years of age. To this end, in 2011, the Ministry of Health increased the quantity of vaccines purchased against hepatitis B by 163% - 83.2 million doses (BRASIL 2010a, 2012a). The vaccine is also available in health establishments for specific groups, regardless of age, such as pregnant women after the first trimester of pregnancy, health workers, firefighters, police officers (military, civil and highway), prisoners (police stations and prisons), hospital and household waste collectors, sexual communicators of HBV carriers, blood donors, men and women who have sexual relations with people of the same sex, lesbians, gays, bisexuals, transvestites and transsexuals, prisoners (prisons, psychiatric hospitals, juvenile institutions, armed forces, among others), manicurists, pedicurists and podiatrists, settlement/encampment populations and indigenous populations (BRASIL 2010b).

The "Brazilian recombinant vaccine", produced by the Butantan Institute, has good immunogenicity (Baldy et al. 2004, Oliveira et al. 2006, Junqueira et al. 2011). From the second half of 2012, the Ministry of Health recommended that the first dose of the monovalent vaccine be administered at birth (hepatitis B), and subsequent doses after 2, 4 and 6 months in its pentavalent form (DTP: diphtheria, tetanus and pertussis, Hib: *Haemophilus influenzae* type B and HB: hepatitis B) (BRASIL 2012b).

Another measure used to prevent HBV infection is passive immunoprophylaxis, through the use of human hyperimmune anti-hepatitis B immunoglobulin (HBIG). Commonly associated with vaccination, HBIG is indicated for use in newborns whose mothers are HBsAg positive, immediately after delivery or up to 12 hours after birth (Shiffman 2010). Other cases indicated for passive prophylaxis are: after accidental exposure to HBsAg-positive blood, after sexual intercourse with HBsAg-positive individuals, patients undergoing liver transplantation, as well as professionals who have had accidents with biological material contaminated with HBV (BRASIL 2006b, CDC 2003, 2007).

1.7 Hepatitis B and Waste Pickers

Recycling waste plays an important role in today's society, as it makes it possible to reuse discarded materials, bringing environmental benefits. Waste pickers collect and separate these materials from the waste and sell them as a source of income and survival (Medeiros & Macêdo, 2006).

In Brazil, the number of waste pickers has grown in recent years, and it is estimated that they already represent 1% of the economically active population, i.e. more than one million people, who optimize their efforts in favor of recycling, and these workers are active in almost every city in the country (CDS-UNB 2005).

In the 1990s, non-governmental institutions supported various meetings of waste pickers throughout Brazil in an effort to strengthen this professional category. In 2001, the "1st National Congress of Waste Pickers" and the "1st Street People's March" took place ([a]) (Godoy 2005, Medeiros & Macêdo 2006, Silva 2007, Costa 2008). These demonstrations enabled waste pickers to organize themselves at regional, state and national level, culminating in the creation of the National Movement of Waste Pickers (MNCR) (Silva 2007, Costa 2008). Thus, in 2002, waste pickers were consolidated as a professional category, regulated and registered in the Brazilian Classification of

Occupations (CBO) (BRASIL 2010c).

According to Miura (2004), the difficulty faced by waste pickers lies not only in the legal recognition of their profession, but also in their right to decent working and living conditions, beyond the strict perspective of survival. The activity of waste picker is seen as a source of dignity and a legitimate way of earning an income. However, the inclusion of these professionals in the job market presents some challenges. It is understood that waste pickers are included because they have a job, but excluded because of the type of work they do, i.e. precarious work, carried out in unhealthy conditions, with a high degree of danger, without social recognition, with often irreversible health risks, and the absence of labor rights (Silva et al. 2002b, Medeiros & Macedo, 2006, Cavalvante & Franco 2007, Dall'Agnol & Fernandes 2007, Gutberlet & Baeder 2008).

Waste pickers have precarious living conditions, living near garbage dumps, landfills or on the outskirts of town, and they collect waste at these sites, in residential areas or in the associations/cooperatives where they work (Godoy 2005, Silva et al. 2005a). Thus, the waste picker lives in a world of marginality and informality, and is not recognized as an agent of environmental transformation (Godoy 2005, Medeiros & Macedo 2006, Costa 2008).

The most frequent agents present in municipal solid waste and in the processes of its management systems, capable of interfering with human health and the environment are: physical agents (odor, dust, vibration of equipment and aesthetic view of the environment), chemical agents (batteries, oils and greases, pesticides/herbicides, solvents, paints, cleaning products, cosmetics, medicines and aerosols) and biological agents (pathogenic microorganisms in paper towels, dressings, disposable diapers, toilet paper, absorbent pads, disposable needles and syringes, condoms, waste from small clinics, pharmacies, laboratories and hospitals mixed with household waste) (Collins & Kennedy 1992, Catapreta & Heller 1999, Ferreira & Anjos 2001, Cavalvante & Franco 2007, Dall'Agnol & Fernandes 2007, Gutberlet & Baeder 2008).

Waste pickers have little knowledge of the risks related to the activity they carry out, and do not have a real view of the risk caused by etiological agents. In this way, waste pickers report information about the adversity of waste picking based on personal experiences, observation of work colleagues and reports of cases experienced by friends or acquaintances who have fallen ill with serious symptoms (Dall'Agnol & Fernandes 2007, Ribeiro & Besen 2007).

The situation of waste pickers is further aggravated by the fact that there is little adherence to the use of PPE, such as appropriate gloves and boots, among others (Gonçalves 2004, Porto et al. 2004, Dall'Agnol & Fernandes 2007, Martins 2007, Ribeiro & Besen 2007, Rozman et al. 2008). Thus, accidents at work are frequent among waste pickers, due to the precariousness and lack of adequate working conditions. These accidents involve injuries from sharp objects from the waste, as well as loss of limbs due to being run over, pressed by compaction equipment and motor vehicles, as well as animal bites (dogs, rats) and insect bites (Poulsen et al. 1995, Ferreira & Anjos 2001, Porto et al. 2004, Ribeiro & Besen 2007, Gutberlet & Baeder 2008).

Therefore, these professionals may be at increased risk of acquiring hepatitis B, since cuts and injuries are caused by the improper disposal of sharps, such as glass, syringes, needles, cans and wood (Velloso et al. 1997, 1998). Porto et al. (2004) found that among the accidents mentioned by waste pickers in a metropolitan landfill in Rio de Janeiro, cuts with glass and punctures with other materials stood out, with hepatitis being one of the diseases mentioned by this population. An investigation into HIV infection among waste pickers in Santos, Sao Paulo, found a high prevalence (34.4%) of HBV infection (Rozman et al. 2007, 2008).

Chapter 2

2. BACKGROUND

Hepatitis B represents a major public health problem. It is estimated that approximately two billion individuals have been exposed to HBV. Of these, 240 million are chronic carriers worldwide and around 600,000 chronically infected individuals die each year from liver complications related to this virus (WHO 2012a). In view of this, research into risk behaviors and estimating the prevalence of hepatitis B in population groups, such as waste pickers, contributes to the epidemiological tracking of this infection.

Accidents involving cuts and punctures due to contact with glass, sharp ferrous materials, contaminated needles and syringes during collection, among others, have been observed among recyclable material collectors (Porto et al. 2004, Rozman et al. 2008, Siqueira & Moraes 2009); this professional category is therefore vulnerable to infection by infectious agents (Porto et al. 2004, Gutberlet & Baeder 2008, Siqueira & Moraes 2009).

However, studies on the health conditions of waste pickers are scarce (Catapreta & Heller 1999, Porto et al. 2004, Cavalcante & Franco 2007, Almeida et al. 2009, Siqueira & Moraes 2009), and the prevalence estimated by Rozman et al. (2007, 2008), in Santos-SP, is the only data in the literature on HBV infection in waste pickers.

In view of the increased risk of these professionals acquiring infections such as hepatitis B and the fact that, to date, there have been few investigations into this virus in waste pickers, this study was proposed with the aim of investigating the epidemiological profile of HBV infection in this population in Goiania. Once we know the associated risk factors, the prevalence in this population, as well as other data such as immunization rates against hepatitis B and occult HBV infection, as well as the circulating viral genotypes, we believe that this study will contribute to the planning and implementation of prevention and control measures aimed   at waste pickers   .

Chapter 3

3. OBJECTIVES
3.1 General Objective
The general aim of this project is to investigate the epidemiological profile of HBV infection in waste pickers in Goiânia, Goiás.

3.2 Specific objectives
- To estimate the prevalence of hepatitis B virus infection among waste pickers in Goiânia-GO;
- Analyze the factors associated with this infection;
- To investigate the occurrence of occult HBV infection in individuals who have been exposed to this virus (anti-HBc reactive);
- To characterize the viral samples, identifying the genotypes circulating in this population;
- To verify the rate of immunization against hepatitis B in the population studied, considering the anti-HBs marker.

Chapter 4

4. METHODOLOGY
Study design, location and population
This research is part of a larger research project called "Investigating viral hepatitis in waste pickers in Goiania-GO, with an emphasis on working conditions".

This is a cross-sectional study carried out in 15 waste picker cooperatives, 12 of which are assisted by the Goiania City Hall and the Social Incubator of the Federal University of Goiás (UFG) and three of which are in the process of being structured, located in Goiania-GO, between April/2010 and May/2011.

There was a prior meeting between the researchers of this study and the leaders of the cooperatives, to clarify the objectives of this research, as well as the criteria for inclusion in the study: to be aged 18 or over and to be linked to one of the cooperatives of waste pickers in Goiania-GO.

The waste pickers were informed about the objectives and procedures of the study, and those eligible (n=432) were invited to take part. Due to illness, only one waste picker refused to take part, so the study population was made up of 431 cooperative waste pickers in Goiania, GO.

This project was approved by the UFG Research Ethics Committee (Protocol No. 002/2010), in Goiania-GO.

4.1 Interview and blood collection
The individuals who consented to take part in the research, by signing an informed consent form (ICF), were interviewed using a standard questionnaire on sociodemographic data (age, gender, *race,* marital status, place of birth, education, family income, length of time working as a waste picker and simultaneous profession as a waste picker) and possible risk factors (use of illicit drugs, presence of a tattoo, history of blood transfusion, surgery, history of imprisonment, unprotected sex with multiple partners, same-sex partners, history of STDs and occupational accidents with sharps) associated with HBV infection, as well as previous vaccination against hepatitis B.

A blood sample (10 mL) was then collected by venipuncture using a disposable syringe and needle. The blood was transported by the team in an appropriate box to the Virology Laboratory (IPTSP/UFG), where it was centrifuged to obtain the sera, which were separated into two aliquots and stored at -20°C (aliquot for serological analysis) and -70°C (aliquot for molecular analysis).

4.3 Serological tests
All samples were tested for the serological markers HBsAg, anti-HBc and anti-HBs by enzyme-linked immunosorbent assay (ELISA) using commercial reagents. The HBsAg-reactive samples were tested for the HBeAg and anti-HBe markers by ELISA.

4.3.1 HBsAg detection
The test performed (Hepanostika HBsAg Ultra, Biomérieux, Germany) consists of an ELISA based on the *sandwich* principle. Briefly, the sample diluent, samples and controls (negative and

positive) were added to the microplate sensitized with monoclonal anti-HBs antibodies in their respective chambers. After incubation, the conjugate (anti-HBs labeled with peroxidase) was added. The plate was incubated again and then washed with phosphate buffer. The substrate (urea peroxide) was then added along with the chromogen (tetramethylbenzidine - TMB). After further incubation, the reaction was stopped by adding 1N sulfuric acid.

According to the manufacturer's instructions, the reaction was read spectrophotometrically at 450 nm. Samples with absorbances greater than or equal to the *cut-off* value, obtained by averaging the absorbances of the negative controls + 0.040, were considered positive.

4.3.2 Detection of total anti-HBc

The principle of the reaction (Hepanostika anti-HBc Uni-Form, Biomérieux) was based on competitive inhibition where each chamber of the plate was coated with HBcAg. The test samples and the negative and positive controls were incubated together with the conjugate (consisting of human anti-HBc antibodies linked to the peroxidase enzyme) and then the plate was washed with the phosphate lid. Subsequently, the substrate/chromogen solution (tetramethylbenzidine/urea peroxide) was added and the reaction incubated. The reaction was stopped with 1N sulfuric acid and the spectrophotometric reading was carried out at 450 nm. The *cut-off* value was obtained using the formula: 0.25 (CNx + 3CPx), where "CNx" is equal to the average absorbance value of the negative controls and "CPx" to that of the positive controls. Thus, positive samples were those with absorbance less than or equal to the *cut-off value*.

4.3.3 Anti-HBs detection

The detection of anti-HBs antibodies was carried out using the qualitative *sandwich* immunoenzymatic method in microplates sensitized with HBsAg (ad and ay), (ETI- AB-AUK-3, Diasorin, Italy). The incubation cap, samples, controls (negative and positive) and calibrators were added to their respective chambers, except for the blank. The plate was incubated and then washed. After adding the conjugate (HBsAg labeled with peroxidase, Tris cap and albumin), the plate was incubated again. After washing, the substrate/chromogen mixture was added and the plate was incubated once more. The reaction was then blocked by sulfuric acid and the spectrophotometric reading (450 nm and 630 nm filters) was carried out.

After validating the test, the *cut-off* was defined by the average absorbance value of calibrator 1 (concentration 10 IU/L). Samples with absorbance values greater than or equal to the *cut-off value* were considered positive.

4.3.4 HBeAg detection

The HBeAg marker was detected by direct non-competitive ELISA using commercial reagents (ETI-EBK PLUS, Diasorin). The incubation cap, samples, calibrator, positive *and* negative controls were added to the microplate sensitized with mouse monoclonal antibodies against the HBV antigen. After this procedure, the plate was incubated and then washed. The diluted enzyme conjugate (peroxidase-conjugated mouse anti-HBe) was added and the plate incubated again. The substrate (hydrogen peroxide) was added along with the chromogen (tetramethylbenzidine - TMB). After incubating again at room temperature, the reaction was stopped by adding 1N sulfuric acid.

According to the manufacturer's instructions, the reaction was read spectrophotometrically at 450/630 nm. Samples with absorbances greater than or equal to the *cut-off* value, obtained by averaging the calibrator values + 0.060 (after subtracting the value of the chamber corresponding to the substrate blank), were considered positive.

4.3.5 Anti-HBe detection

The anti-HBe marker was detected by competitive ELISA using commercial reagents (ETI-AB-EBK PLUS, Diasorin). In the microplate sensitized with mouse monoclonal antibodies against the HBV "e" antigen, the incubation cap, samples, positive and negative controls, calibrator and a solution containing recombinant HBeAg (the antibodies in the sample compete with the monoclonal antibodies for HBeAg) were added in all chambers. After this procedure, the plate was incubated and washed. Subsequently, the enzyme conjugate (peroxidase-conjugated mouse anti-HBe) was added and the plate was incubated again. The plate was then washed and the chromogen/substrate added. After further incubation at room temperature, the reaction was stopped by adding 1N sulfuric acid.

In accordance with the manufacturer's instructions, the reaction was read spectrophotometrically using a 450/630 nm double filter. Samples with absorbances greater than the *cut-off* value, obtained by averaging the calibrator values + 0.500 (after subtracting the value of the chamber corresponding to the substrate blank), were considered positive.

4.4 Molecular Tests

4.4.1 HBV-DNA extraction

The HBsAg and/or anti-HBc reagent samples were submitted to viral nucleic acid extraction, carried out in two stages (Niel et al. 1994). The first consisted of adding 250 yL of serum to 80 yL of lysis solution, which was made up of solution A (200 mM Tris, 1% SDS, 700 mM NaCl, 20 mM EDTA and 0.1 mg/mL tRNA) and solution B (2 mg/mL proteinase K dissolved in 0.36 mM CaCl2 solution; the final pH was adjusted to 9.0 using HCl). The sera, together with the lysis solution, were incubated for 4 hours at 37°C. The viral DNA was extracted using phenol/chloroform, centrifuged, and the upper phase transferred to tubes containing ethanol. These were kept at -20°C overnight.

The second stage of the reaction was characterized by centrifugation of the material and the formation of a precipitate, which was then washed with 70% ethanol, dried and resuspended in 30 yL of milliQ (auloclaved) water.

4.4.2 Polymerase chain reaction (PCR)

After extracting the DNA, it was submitted to a *semi-nested* PCR for amplification of the pre-S/S region, which was carried out in two stages (Motta-Castro et al. 2005).

For the first, called PCR-1, a mixture was used containing 1 pmol of the PS1 and S2/S22 *primers* (Table 2) (Invitrogen), responsible for amplifying the DNA fragment from nucleotide 2826 to 841 of the genome; 0.2 mM of each dNTP (dATP, dTTP, dCTP, dGTP), 3 mM of magnesium chloride ($MgCl2$) and 1 U of Taq DNA polymerase, in a final volume of 50 yl. The HBV-DNA from each sample extracted in the previous phase (2 yl.) was mixed with the components described above, and this mixture was taken to the thermal cycler following the program: 94°C for 3 minutes (initial denaturation), 30 cycles of 95°C for 30 seconds, 52°C for 40 seconds and 72°C for 2 minutes (denaturation, annealing and extension, respectively) and 72°C for 7 minutes (final elongation). The expected final product of this reaction was 1236 base pairs (bp).

The second stage (PCR-2) was carried out using *primers* PS1 and SR (Table 2) (Invitrogen), as well as the same components mentioned above. The PCR-1 product was added to the new reaction mixture and taken to the thermal cycler, following the program: 94°C for 3 minutes (initial denaturation), 30 cycles of 94°C for 20 seconds, 55°C for 20 seconds and 72°C for 1 minute (denaturation, annealing and extension, respectively) and 72°C for 7 minutes (final elongation). The expected PCR-2 product was 1099 base pairs (bp).

Table 2 - Sequence of the *primers* used in PCR-1 and PCR-2

Primers	*Sense*	Sequence (5' ^ 3')
PS1	*Sense*	CCA TAT TCT TGG GAA CAA GA
S2	*Anti-sense*	GGG TTT AAA TGT ATA CCC AAA GA
S22	*Anti-sense*	GTA TTT AAA TGG ATA CCC ACA GA
SR	*Anti-sense*	CGA ACC ACT GAA CAA ATG GC

4.4.3 Agarose gel electrophoresis

Ethidium bromide was added to the agarose gel prepared at 2% in TBE (Tris-Borate EDTA) buffer at a final concentration of 0.5 iig/ml. The products obtained during PCR-2 and the molecular weight marker (100 bp - Invitrogen) were mixed with bromophenol blue dye, applied to the gel and subjected to electrophoresis. The bands formed during the run were visualized using ultraviolet light in a transluminator.

4.5 Nucleotide Sequencing
4.5.1 Amplified samples

Reactive HBV-DNA samples were re-amplified in the S region by *semi-nested* PCR for subsequent sequencing (Motta-Castro et al.
2008). The first PCR used external *primers* PS1, S2 and S22, under the same conditions described in PCR-1, with a final volume of 50 |iL. However, the second PCR was carried out with *primers* S1, S2 and S22 (Table 4), with a final volume of 100 |iL. For amplification, the samples were sent to the thermal cycler with the following program: 94°C for 3 minutes for initial denaturation, 30 cycles of (95°C for 30 seconds, 52°C for 40 seconds, 72°C for 2 minutes), ending with elongation at 72°C for 7 minutes.

4.5.2 Purification of PCR products

Five volumes of PBI cap (ligation cap) were added to one volume of PCR product and homogenized in a vortex. The sample was then placed on the QIAquick column to bind the DNA, which was centrifuged (approximately 13,000 rpm) for 30-60 seconds and the liquid from the collection tube was discarded. The column was placed back in the tube, 0.75 mL of PE cap (wash cap) was added and centrifuged for 30-60 seconds. Finally, the liquid from the collection tube was discarded and the column was repositioned in the tube and centrifuged for 1 minute.

The QIAquick column was placed in a clean 1.5 mL microcentrifuge tube, and 30 LITERS of EB cap (elution cap) were added to the center of the membrane, left to stand for 1 minute, centrifuged again, and the DNA stored at - 20°C.

4.5.3 S region sequencing

The S region of the HBV genome was sequenced using the *BigDye Terminator version 3.01 Cycle Sequencing Kit (Applied Biosystems, Foster City, CA, USA)* and the specific *primers* S1, S4, S7, S2 and S22. The reaction was pre-mixed in a labeled tube according to Tables 3 and 4.

Table 3 - Reagents used in the reaction premix

Pre-mixing the reaction	Quantity for 1 sample
Water	6 |1L
5X sequencing cap	3 |1L
Primer 2 pmol	3 |1L
BigDye	1 |1L
TOTAL	13 jiL

Table 4 - *Primers* for sequencing

Primer	Position nt*	Sequence (5' ^ 3')

S1 (*sense*)	nt124-143	CTT CTC GAG GAC TGG GGA CC
S2 (*anti-sense*)	nt841-819	GGG TTT AAA TGT ATA CCC AAA GA
S22 (*anti-sense*)	nt841-819	GTA TTT AAA TGG ATA CCC ACA GA
S4 (*sense*)	nt416-436	TGC TGC TAT GCC TCA TCT TCT
S7 (*anti-sense*)	nt 676-656	TGA GCC AGG AGA AAC GGG CT

*The nucleotides (nt) are numbered at the cleavage site (EcoRI), located in the pre-S2 region (complete genome 3221pb).

13 |iL of the pre-mix were distributed on the sequencing plate and 2 |iL of pure DNA sample in a dark environment and centrifuged (700 rpm for 1 min). The capped plate was incubated in the thermal cycler according to the program described below.

	95°C for 20 sec
25 Cycles	50°C for 15 sec
	60°C for 60 sec
	4°C - finish

After this stage, the samples were precipitated by adding 60 |iL of 65% Isopropanol (Merck) to each pogo stick. It was then homogenized in a vortex (1 min), waiting 20 minutes at room temperature in the dark. Afterwards, it was centrifuged for 45 minutes (4000 rpm) at 20°C. Afterwards, the supernatant was discarded and 250 ul. of 60% ethanol was added, which was centrifuged for 10 min (4000 rpm) at 20°C. The supernatant was discarded again and 100 lL of 60% ethanol was added, centrifuged for 10 minutes (4000 rpm) at 20°C. The supernatant was discarded and a final centrifugation was carried out for 1 minute (500 rpm).

The plate was taken to the thermal cycler on the Drying program (95°C for 2 min). Then, 10 lL of Hi Di formamide (Applied Biosystems) was added to each chamber, and the plate was incubated in the thermal cycler on the Denature program (5 min at 95°C). The plate was removed from the thermal cycler, incubated on ice for 2 minutes and centrifuged (700 rpm for 1 min). Finally, the plate was placed in the automatic sequencer (ABI 3130, Applied Biosystems) to read the electropherograms.

4.5.4 Sequence analysis

Based on the analysis of the sequences obtained, a consensus sequence of the *sense* and *antisense* strands was assembled using the SeqMan II version 5.01 program (DNASTAR). The genotypes were identified by aligning the consensus sequences of each sample using the Clustal X program (Thompson et al. 1997) together with representative sequences of HBV genotypes (A-H) obtained from GenBank (http://www.ncbi.nlm.nih.gov/). The alignment was edited using the BioEdit program (version 7.1.3 - Ibis Biosciences, USA).

The phylogenetic tree was constructed using the MEGA program (Molecular Evolutionary Genetics Analysis) version 4 (Tamura et al. 2007). The analyses were carried out using the *Neighbor-Joining* method, the Kimura 2-parameter nucleotide substitution model, and the robustness of the phylogenetic groups was evaluated using a *bootstrap* of 1025 repeats.

4.6 Data Analysis

The data collected during the interviews, as well as the results of the serological and molecular tests, were analyzed using EpiInfo verse 6.04 (*Centers for Disease Control and Prevention,* Atlanta, GA). Prevalence and odds estimates were calculated with a 95% confidence interval. Factors associated with HBV exposure (HBsAg and/or anti-HBc positivity) by univariate analysis (p<0.10) were further analyzed by logistic regression using SPSS version 11.0 *for Windows*. Values of p<0.05 were considered statistically significant. The chi-square, chi-square for trend and Fisher's exact tests were used when appropriate.

Chapter 5

5. RESULTS

5.1 Characteristics of the population studied

The sociodemographic characteristics of 431 waste pickers in Goiânia-GO are shown in Table 1. The average age of this population was 36.9 years (SD 13.6). There was a predominance of females, accounting for 62.4% of the population. With regard to race, 49.9% reported being brown, 28.8% black, 20.6% white and 0.7% yellow.

With regard to place of birth, 41.6% were from Goiás, while 58.4% reported having been born in other states of the Federation. When stratifying the population according to marital status, 48.7% of the interviewees reported being married or in a consensual union, 39% single and 12.3% widowed or separated.

Almost 30% of those interviewed reported less than four years of schooling, 39.5% between four and eight years and 31% more than eight years. Of the total, 51.0% reported an income of between one and two minimum wages, 43.0% less than one minimum wage and 6.0% between three and five minimum wages. In addition, 52.7% had worked as waste pickers for one year or less, 36.4% for between two and 10 years and 10.9% for more than 10 years.

Table 1 - Sociodemographic characteristics of 431 waste pickers in Goiânia, Goiás

Features	N	%
Mean age (SD) 36.9 years (13.6)		
Sex		
Female	269	62,4
Male	162	37,6
Race		
White	89	20,6
Black	124	28,8

Brown	215	49,9
Yellow	3	0,7
Naturalness		
Goiás	178	41,6
Other state	250	58,4
No information 3		
Marital status		
Single	168	39,0
Married/consensual union	210	48,7
Widowed/ Separated	53	12,3
No information 3		
Schooling (years)		

<4	127	29,5
4-8	170	39,5
>8	133	31,0
No information 1		
Average schooling (6.1 years)		
Family income		
< 1 SM *	185	43,0
1-2 SM	220	51,0
3-5 SM	26	6,0
Time working as a waste picker (years)		
< 1	227	52,7
2-10	157	36,4

>10	47	10,9
Average working time (3.95 years)		

SD: Standard Deviation; * Minimum Wage

5.2 HBV serological markers in waste pickers

The anti-HBc marker was detected in isolation in 2.3% of the individuals studied. This same marker associated with HBsAg was reactive in 0.7% and, when associated with anti-HBs, it was present in 9.7% of the collectors (Table 2), resulting in an overall prevalence of 12.8% (95% CI: 9.8-16.2) for HBV infection. In 53 (12.3%) waste pickers, there was isolated positivity for the anti-HBs marker, indicating prior vaccination against hepatitis B. Susceptibility to HBV, characterized by the absence of any marker for hepatitis B, was observed in 323 (74.9%) waste pickers.

Table 2 - Prevalence of serological markers for hepatitis B in 431 waste pickers from Goiania, Goiás.

Category	Markers	Positive		95% CI
		N	%	
Infected	anti-HBc	10	2,3	1,2- 4,4
	anti-HBc/HBsAg	3	0,7	0,2 - 2,2
	anti-HBc/anti-HBs	42	9,7	7,2 - 12,8
	Global	55	12,8	9,8 - 16,2
Immune	anti-HBs	53	12,3	9,4 - 15,7
Susceptible	no marker	323	74,9	70,5 - 78,9

CI: confidence interval

5.3 Risk factors associated with HBV infection

Table 3 shows the univariate analysis of the risk factors associated with HBV infection, with their respective 95% confidence intervals, in waste pickers from Goiânia-GO. Among the factors analyzed, those that were statistically significant in the univariate analysis were: age, schooling, history of imprisonment and STDs. Use of illicit drugs (p=0.08) and unprotected sex with multiple partners (p=0.05) showed a marginal association.

After multivariate analysis (Table 4), there was a greater chance of acquiring HBV infection with increasing age. Thus, waste pickers aged between 41 and 50 were 8.9 (95% CI: 2.4-32.1) times more likely, those aged 51-60 were 17.8 (95% CI: 4.6-58.0) times more likely and those over 60 were 38.5 (95% CI: 8.5-174.7) times more likely to acquire HBV than those aged 30 or under. As for the use of illicit drugs, waste pickers who reported being users were 3.1 (95% CI: 1.2-8.0) times more likely to acquire HBV than non-users.

Table 3 - Risk factors associated with HBV infection in waste pickers recyclables in Goiania-GO

Risk factor	HBV Positive / Total [b]	%	Chance estimate (95% CI)[a]	P
Age (years)				
< 30	4/127	3,1	1,0	
31-40	10/107	9,3	3,2 (0,96-10,4)	0,04
41-50	13/68	19,1	7,3 (2,3- 23,3)	< 0,01
51-60	18/55	32,7	15,0 (4,8-47,0)	< 0,01
> 60	10/21	47,6	28,0 (7,5- 103,9)	< 0,01
Sex				
Female	30/235	12,8	1, 0	
Male	25/143	17,5	1,4 (0,8-2,6)	0,21

Schooling (years)				
>8	7/103	6,8	1,0	
4-8	19/155	12,2	1,9 (0,8-4,7)	0,15
<4	29/119	24,4	4,4 (1,8-10,6)	0,01
Time working as a waste picker (years)				
< 1	27/195	13,8	1,0	
2-10	20/140	14,3	1,0 (0,5-2,0)	0,90
>10	8/43	18,6	1,4 (0,5-3,6)	0,42
Accident with a sharp object from the garbage				
No	28/197	14,2	1,0	
Yes	27/179	15,1	1,1 (0,6-2,0)	0,80
Blood transfusion				

No	46/329	14,0	1,0	
Yes	9/48	18,8	1,4 (0,6-3,1)	0,38
Surgery				
No	20/174	11,5	1,0	
Yes	35/204	17,2	1,6 (0,9-2,9)	0,12
Tattoo				
No	43/307	14,0	1,0	
Yes	12/71	16,9	1,2 (0,6-2,5)	0,53
Use of illicit drugs				
No	40/307	13,0	1,0	
Yes	15/71	21,1	1,8 (0,9-3,5)	0,08
Previous arrests				

No	38/310	12,3	1,0	
Yes	16/65	24,6	2,3 (1,2-4,5)	0,01
Unprotected sex with multiple partners				
No	26/223	11,7	1,0	
Yes	29/155	18,7	1,7 (1,0-3,1)	0,05
Same-sex partner				
No	50/338	14,8	1,0	
Yes	4/22	18,2	1,3 (0,4-3,9)	0,66
History of STDs [c]				
No	35/286	12,2	1,0	
Yes	20/78	25,6	2,5 (1,3-4,6)	0,03

[a]CI: confidence interval;[b] The denominator reflects the number of waste pickers who answered the question. Those vaccinated against hepatitis B (anti-HBs isolate) were excluded; [c]STD: sexually transmitted disease.

Table 4 - Multivariate analysis of risk factors associated with HBV infection in waste pickers in Goiania-GO

Risk factors	Chance estimate (95% CI)[a]		P
	Not adjusted	Adjusted [b]	
Age (years)			
< 30	1,0	1,0	
31-40	3,2 (0,96-10,4)	3,5 (1,0-12,4)	0,05
41-50	7,3 (2,3- 23,3)	8,9 (2,4-32,1)	0,00
51-60	15,0 (4,8-47,0)	17,8 (4,6-68,0)	0,00
> 60	28,0 (7,5- 103,9)	38,5 (8,5-174,7)	0,00
Sex Female Male	1,0 1,4 (0,8-2,6)	1,0 0,8 (0,4-1,9)	0,66
Schooling (years)			

>8	1,0	1,0	
4-8	1,9 (0,8-4,7)	2,0 (0,7-5,3)	0,18
<4	4,4 (1,8-10,6)	2,1 (0,8-5,6)	0,15
Use of illicit drugs			
No	1,0	1,0	
Yes	1,8 (0,9-3,5)	3,1 (1,2-8,0)	0,02
Previous arrests			
No	1,0	1,0	
Yes	2,3 (1,2-4,5)	1,4 (0,5-3,4)	0,50
Unprotected sex with multiple partners			

No	1,0	1,0	
Yes	1,7 (1,0-3,1)	1,4 (0,6-3,2)	0,38
History of STDs			
No	1,0	1,0	
Yes	2,5 (1,3-4,6)	1,4 (0,7-3,0)	0,35

[a]CI: confidence interval; bChance estimate adjusted for: age, sex, education, illicit drug use, history of imprisonment, unprotected sex with multiple partners and history of sexually transmitted disease (STD)

5.4 Detection of HBeAg/anti-HBe markers, HBV-DNA and risk characteristics of HBsAg-positive individuals

The three HBsAg-positive collectors (Table 5) were anti-HBe reactive. HBV-DNA was detected in two of them (CT- 43 and CT- 129).

Table 5 - HBeAg, anti-HBe and HBV-DNA markers in HBsAg reactive waste pickers

Collectors	HBeAg	Anti-HBe	HBV-DNA
CT- 43	-	+	+
CT- 90	-	+	-
CT- 129	-	+	+

These three waste pickers were male. The first (CT-43) was 40 years old and reported a sharps accident, multiple sexual partners (n=10) and a history of STDs. The second (CT-90) was 30 years old and had a history of a sharps accident, illicit drug use, a tattoo, a history of imprisonment and multiple sexual partners (n=10). The last (CT-129), aged 52, reported illicit drug use, a history of imprisonment and multiple sexual partners (n>100).

5.5 Hidden hepatitis B virus infection

The isolated anti-HBc and anti-HBc/anti-HBs reactive samples were subjected to HBV-DNA testing to determine the presence of occult HBV infection. One sample (CT-160, anti-HBc/anti-HBs reactive) was positive for viral DNA, resulting in an occult infection rate of 1.9% (1/52) in anti-HBc reactive waste pickers in Goiânia-GO.

The individual with occult HBV infection was female, aged 42, and reported an accident with sharps during scavenging.

5.6 Molecular characterization of HBV-DNA positive samples

According to the phylogenetic tree of the HBV S-region (Figure 12), the three HBV-DNA positive samples were characterized as genotypes A (subgenotype A1), D (subgenotype D3) and F (subgenotype F2).

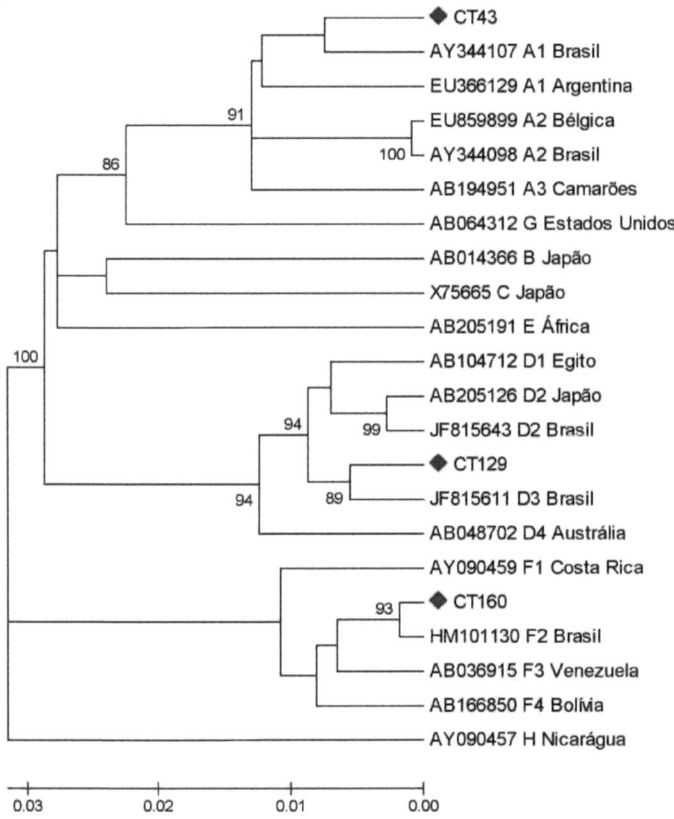

Figure 12 - Phylogenetic tree of the S region of HBV, including the 3 isolates from waste pickers in Goiania-GO and 19 sequences from GenBank (accession number and country of origin are indicated)

Chapter 6

6. DISCUSSION

This study is the first seroepidemiological and molecular investigation of hepatitis B virus infection in waste pickers. Although this study was carried out in cooperatives/associations responsible for the collection and separation of recyclable materials, and therefore does not represent the entire population of waste pickers, the information obtained in this study can support the development of strategies for the prevention and control of hepatitis B aimed at this population.

After analyzing the sociodemographic characteristics, we found that the average age of 36.9 years (SD 13.6) was close to that found in Duque de Caxias-RJ (44 years) (Porto et al. 2004) and Santos-SP (42.4 years) (Rozman et al. 2008). The majority of the population studied was female. This is in line with other studies carried out on waste pickers by Porto et al. (2004), in Duque de Caxias-RJ, and Almeida et al. (2009), in Governador Valadares-MG. However, this diverges from other investigations conducted in Brazil by Silva et al. (2005a), in Pelotas-RS, and Rozman et al. (2008), in Santos-SP. According to Porto et al. (2004), women are more likely to work in the cooperatives on the grounds that working on the sorting lines (mechanical conveyors operated by waste pickers linked to the cooperative) requires less physical effort than on the ramp (open-air areas where the trucks deposit the waste to be spread out and covered with soil by the tractors). This may explain the predominance of women in waste picker cooperatives in Goiânia.

Half of the population studied was brown (49.9%). This is similar to the data shown in the study by Silva et al. (2005a), in which 47.5% of waste pickers were brown. In 2010, according to the IBGE Census (2010) (Brazilian Institute of Geography and Statistics), Brazil had a population of 191 million inhabitants, of which 91 million self-declared as white (47.7%), 15 million as black (7.6%), 82 million as brown (43.1%), 2 million as yellow (1.1%) and 817 thousand indigenous (0.4%). Thus, the rate of brown individuals in this study was close to that found in the Brazilian population. On the other hand, the percentage of white waste pickers (20.6%) was approximately half that observed by the IBGE Census (2010), while the percentage of black waste pickers (28.8%) was almost four times higher.

With regard to schooling, the majority of the population (69%) reported up to eight years of study, with an average of 6.1 years. Few years of schooling were also found in this same population by other authors (Porto et al. 2004, Silva et al. 2005a, Alvarado-Esquivel et al. 2008, Rozman et al. 2008). According to the IBGE (2010), the average schooling of the population over 10 years of age was 7.2 years in 2010, slightly higher than that observed in this study. The low or average schooling of waste pickers can be a factor in their exclusion from the labor market (Silva 2002b, Magera 2003). The few years or lack of schooling, combined with the condition of living from scavenging, shows that, for many workers, this activity is the only way to guarantee survival and the possibility of inclusion in the job market (Medeiros & Macedo 2006).

Almost half of the population reported a family income of less than one minimum wage (43.0%). The low and unstable income of waste pickers has also been shown in other studies (Gongalves 2004, Porto et al. 2004, Silva et al. 2005a, Ribeiro & Besen 2007, Bossi 2008, Rozman et al. 2008, Kirchner et al. 2009). According to the IBGE (2010), the average monthly income of permanent private households was R$ 2,222.00, reaching R$ 2,407.00 and R$ 1,051.00 in urban and rural areas, respectively. These figures were higher than those reported by the individuals studied, thus indicating that waste pickers are a low-income population.

In addition, the average working time was 3.95 years, slightly higher than that found by Silva et al. (2005a) (3 years) and lower than that found by Rozman et al. (2008) (6.52 years). The Goiania Selective Collection Program, founded in 2008, has made it possible for all the recyclable materials collected by the City Hall to be donated to the cooperatives and associations of waste pickers that have signed up to the Program, which has meant an increase in the number of cooperatives/associations in recent years, thus attracting new workers to this activity in Goiania (Goiania City Hall 2009), a fact that may justify the short time spent working in the cooperatives reported by almost half of the population in this study. Furthermore, Ferreira (2005) considers waste picking and sorting to be a recent form of income for workers, especially after globalization. After all, these workers didn't have the necessary skills to meet the demands of the new market and became waste pickers.

The overall prevalence of HBV infection estimated in this study of 12.8% (95% CI: 9.8-16.2) was higher than that reported in a Brazilian population-based survey of 5.3% (95% CI: 4.6-6.1) (Pereira et al. 2009). However, it was lower than that observed in a study of 250 waste pickers by Rozman et al. (2008) of 34.4% (95% CI: 28.5-40.2). This discrepancy can be explained by the difference between the population in this study and the waste pickers in Santos-SP, who were older and predominantly men. The average working time of waste pickers in Santos-SP was longer than in Goiânia-GO. In addition, risk behaviors related to parenteral transmission (10% of waste pickers reported injecting drug use) and sexual transmission (50% reported more than 10 sexual partners in their lifetime) were highlighted by Rozman et al. (2008). Furthermore, Santos-SP is a port city, and this environment can favour the marginalization of socially excluded populations such as waste pickers (Lacerda et al. 1996, Etzel et al. 2001).

Among the risk factors associated with HBV infection, there was a greater chance of acquiring the virus with increasing age. Exposure to HBV is associated with risk behaviors acquired throughout life, both related to sexual transmission and the use of illicit drugs (Bernabe-Ortiz et al. 2011, Matos et al. 2012). With regard to illicit drug use, waste pickers who reported being users were 3.1 (95% CI: 1.2-8.0) times more likely to have acquired HBV than non-users. The use of illicit drugs has also been described as a risk factor for acquiring HBV in other studies (Oliveira et al. 2005, Sutton et al. 2006, Reimer et al. 2007, Tseng et al. 2007, Amesty et al. 2008, Ferreira et al. 2009).

As mentioned, the sexual route has been identified as one of the main ways of transmitting HBV (Alter 2003, Kao 2011). In the univariate analysis, the variable history of STDs (p=0.03) was associated with HBV infection, and unprotected sex with multiple partners was marginally associated (p=0.05). These factors were associated with this infection in other studies carried out in Brazil (Lewis- Ximenez et al. 2002, Ferreira et al. 2009, Pinho et al. 2011, Matos et al 2012).

The prison history variable was also associated with HBV infection in the univariate analysis (p=0.01). Prison is characterized as a promiscuous environment, with poor hygiene habits and a marginalized population (murderers, drug dealers and injecting drug users) (Catalan-Soares et al. 2000, Adjei et al. 2008, Hennessey et al. 2008, Stief et al. 2010, Saiz de la Hoya et al. 2011), conditions which favor the spread of HBV, as well as other infectious agents (Miranda et al. 2004, Strazza et al. 2007, Adjei et al. 2008, Nelwan et al. 2010, Keramat et al. 2011, Mir-Nasseri et al. 2011).

Accidents with sharp objects from the garbage, although not associated with HBV infection, were reported by 47.6% (179/376) of the waste pickers. The biological risk involved in waste picking, coupled with low schooling and inexperience in handling potentially contaminated materials, reveals the importance of health education activities for the waste picker population. The correct segregation of each type of waste and investments in occupational safety, with the provision and training in the use of individual and collective protection equipment, are important measures for the quality of life of waste pickers (Cavalvanti & Franco 2007).

In Brazil, given the diversity of the population and the prevalence of hepatitis B, the identification of circulating genotypes is of great importance, as it contributes to knowledge of the epidemiology of HBV in the country. In this study, genotypes A (subgenotype A1), D (subgenotype D3) and F (subgenotype F2) were identified, which is in agreement with other investigations carried out in the country (Motta-Castro et al. 2005, 2008, Paraná & Almeida 2005, Mello et al. 2007, Sitnik et al. 2007, Bottecchia et al. 2008, Matos et al. 2009, Santos et al. 2010).

HBV-DNA has been detected in the absence of the HBsAg marker, a profile known as occult infection (Ocana et al. 2011, Candotti et al. 2012). In this study, the prevalence of occult infection of 1.9% in anti-HBc reactive collectors was similar to that observed by Matos et al. (2009) (1.7%) in an Afro-descendant community in Goiás, using the same serological and molecular methods. It is worth highlighting the possibility of hepatitis B transmission, even in the absence of the HBsAg marker, in patients with occult infection (Conjeevaram & Lok 2001, Raimondo et al. 2007, Hollinger & Sood

2010, Larrubia et al. 2011, Raimondo 2012).

In Brazil, vaccination against hepatitis B was introduced into the National Immunization Program (PNI) in 1988 and is currently available in public health services for specific age groups, such as children and adults under 29 years of age, as well as for individuals with a greater chance of exposure or in cases of accidents with contaminated materials (BRASIL 2010a, BRASIL 2010b). However, not specifically for waste pickers. In this study, a high rate of susceptibility to this infection was observed, as well as the anti-HBs marker being detected in isolation in only 12.3% of the population studied, suggesting a low rate of immunization against hepatitis B. In fact, only 29.8% (28/94) of the waste pickers, who were in the age group in which the vaccine was made available by the Ministry of Health (up to 24 years old), had been previously immunized.

This study shows a high prevalence of HBV when compared to the urban population of the Midwest region (Pereira et al. 2009) and, on the other hand, a low rate of immunization against hepatitis B. These data, combined with risk behaviors, demonstrate the need for effective interventions aimed at preventing and controlling hepatitis B in waste pickers, such as hepatitis B vaccination in places where this population is concentrated, including cooperatives. Such actions will lead to a better quality of life and care for these individuals.

CONCLUSIONS

♦ The overall prevalence of hepatitis B virus infection among waste pickers in Goiania-GO of 12.8% is higher than that found in the population-based survey carried out in our region;

♦ The factors associated with this onset, age greater than or equal to 40 and illicit drug use, suggest sexual and parenteral transmission of HBV;

♦ The presence of occult HBV infection in anti-HBc reactive individuals, with a rate of 1.9%, shows the possibility of viral transmission, even in the absence of the HBsAg marker;

♦ HBV genotypes A (subgenotype A1), D (subgenotype D3) and F (subgenotype F2) were identified in the infected individuals, corroborating data from other studies published in Brazil;

♦ Only 12.3% of the population was positive for the anti-HBs marker alone, thus indicating a low rate of immunization against hepatitis B among waste pickers in Goiania-GO.

FINAL CONSIDERATIONS

After the tests were *carried out* at the IPTSP/UFG Virology Laboratory, the serological results were delivered to the cooperatives gradually. We advised them on the importance of vaccination against hepatitis B, and the individuals who tested positive for HBsAg were advised and referred to the Testing and Counseling Center (CTA) of the Municipal Health Department of Goiania-GO.

In addition, an extension project was carried out in the recyclable material collectors' cooperatives, enabling health education activities, with an emphasis on the use of PPE. All the cooperative waste pickers were given a PPE kit, including a mask, goggles, gloves and boots, with the aim of encouraging them to use the equipment properly.

The development of this study has awakened the need for environmental awareness, with incentives for waste management, as well as strengthening the selective collection program, and consequently the cooperatives of recyclable material collectors. Finally, we can't fail to highlight the lessons learned about the difficulties and stigmas faced by waste pickers in favor of recognizing their professional category, which is fundamental in society, since it makes it possible to reuse discarded materials, bringing environmental benefits.

REFERENCES

Abdou Chekaraou M, Brichler S, Mansour W, Le Gal F, Garba A, Dény P, Gordien E 2010. A novel hepatitis B virus (HBV) subgenotype D (D8) strain, resulting from recombination between genotypes D and E, is circulating in Niger along with HBV/E strains. J Gen Virol 91(6): 1609-20.

Adjei AA, Armah HB, Gbagbo F, Ampofo WK, Boamah I, Adugyamfi C, Asare I, Hesse IFA, Mensah G 2008. Correlates of HIV, HBV, HCV and syphilis infections among prison inmates and officers in Ghana: A national multicenter study. BMC Infect Dis 8(33): 1-12.

Aires RS, Matos MA, Lopes CL, Teles SA, Kozlowski AG, Silva AM, Filho JA, Lago BV, Mello FC, Martins RM 2012. Prevalence of hepatitis B virus infection among tuberculosis patients with or without HIV in Goiânia City, Brazil. J Clin Virol 54(4): 327-31.

Alavian SM, Miri SM, Jazayeri SM 2012. Hepatitis B vaccine: prophylactic, therapeutic, and diagnostic dilemma. Minerva Gastroenterol Dietol 58(2): 167-78.

Almeida JD, Zuckerman AJ, Taylor PE, Waterson AP 1969. Immune electron microscopy of the Australia - SH (serum hepatitis) antigen. Microbes 1: 117-123.

Almeida JR, Elias ET, Magalhaes MA, Vieira AJD 2009. Effect of age on the quality of life and health of waste pickers from an association in Governador Valadares, Minas Gerais, Brazil. Ciênc Saúde Coletiva 14(6): 21692179.

Alter HJ, Seeff LB, Kaplan PM, McAuliffe VJ, Wright EC, Gerin JL, Purcell RH, Holland PV, Zimmerman HJ 1976. Type B hepatitis: the infectivity of blood positive for e antigen and DNA polymerase after accidental needlestick exposure. N Engl J Med 295(17): 909-13.

Alter MJ 2003. Epidemiology and prevention of hepatitis B. Semin Liver Dis 23 (1): 39-46.

Alvarado-Esquivel C, Liesenfeld O, Márquez-Conde JA, Cisneros-Camacho A, Estrada-Martínez S, Martínez-García SA, González-Herrera A, García-Corral N 2008. Seroepidemiology of infection with Toxoplasma gondii in waste pickers and waste workers in Durango, Mexico. Zoonoses Public Health 55(6): 306-312.

Alvarado-Mora MV, Botelho L, Gomes-Gouvêa MS, de Souza VF, Nascimento MC, Pannuti CS, Carrilho FJ, Pinho JR 2011. Detection of Hepatitis B virus subgenotype A1 in a Quilombo community from Maranhao, Brazil. Virol J 8:415.

Amesty S. Ompad DC, Galea S, Fuller CM, Wu Y, Koblin B, Vlahov D 2008. Prevalence and correlates of previous hepatitis B vaccination and infection among young drug-users in New York City. J Community Health 33(3): 139-48.

Arauz-Ruiz P, Norder H, Robertson BH, Magnius LO 2002. Genotype H: a new Amerindian genotype of hepatitis B virus revealed in Central America. J Gen Virol 83(Pt8): 2059-2073.

Arbuthnot P, Kew M 2001. Hepatitis B virus and hepatocellular carcinoma. Int J Exp Pathol 82(2): 77-100.

Aspinall EJ, Hawkins G, Fraser A, Hutchinson SJ, Goldberg D 2011. Hepatitis B prevention, diagnosis, treatment and care: a review. Occup Med (Lond) 61(8):531-40.

Azevedo MSP, Cardoso DDP, Martins RMB, Daher RR, Camarota SCT, Barbosa AJ 1994. Serologic screening for hepatitis B in health professionals in the city of Goiânia - Goias. Rev Soc Bras Med Trop 27(3): 157- 62.

Badur S, Akgün A 2001. Diagnosis of hepatitis B infections and monitoring of treatment. J Clin Virol 21(3): 229-237.

Baffis V, Shrier I, Sherker AH, Szilagyi A 1999. Use of interferon for prevention of hepatocellular carcinoma in cirrhotic patients with hepatitis B or hepatitis C virus infection. Ann Intern Med 131(9): 696-701.

Balayan MS, Andjaparidze AG, Savinskaya SS, Ketiladze ES, Braginsky DM, Savinov AP, Poleschuk VF 1983. Evidence for a virus in non-A, non-B hepatitis transmitted via the fecal-oral route. Intervirology 20(1): 23-31.

Baldy JL, de Lima GZ, Morimoto HK, Reiche EM, Matsuo T, de Mattos ED, Sudan LC 2004. Immunogenicity of three recombinant hepatitis B vaccines administered to students in three doses containing half the antigen amount routinely used for adult vaccination. Rev Inst Med Trop 46(2): 103-7.

Bancroft WH, Mundon FK, Russell PK 1972. Detection of additional antigenic determinants of hepatitis B antigen. J Immunol 109(4): 842-848.

Bastos CC 2000. Seroepidemiological study of hepatitis B, hepatitis C and human immunodeficiency virus in patients with sickle cell disease. Master's dissertation - Institute of Tropical Pathology and Public Health/Federal University of Goiás (IPTSP/UFG) 79 p.

Baumert TF, Rogers AS, Hasegawa K, Liang TJ 1996. Two core promoter mutations identified in a hepatitis B virus strain associated with fulminant hepatitis result in enhanced viral replication. J Clin Invest 98(10): 2268-2276.

Bayer ME, Blumberg BS, Werner B 1968. Particles associated with Australia antigen in the sera of patients with leukemia, Down's syndrome and hepatitis. Nature 218(146): 1057-1059.

Beck J, Nassal M 2007. Hepatitis B virus replication. World J Gastroenterol 13(1): 4864.

Beckebaum S, Sotiropolos GC, Gerken G, Cinnati VR 2009. Hepatitis B and liver transplantation: 2008 update. Rev Med Virol 19(1): 7-29.

Befeler AS, Di Bisceglie AM 2000. Hepatitis B. Infect Dis Clin North Am 14(3): 617632.

Bernabe-Ortiz A, Carcamo CP, Scott JD, Hughes JP, Garcia PJ, Holmes KK, 2011. HBV Infection in Relation to Consistent Condom Use: A Population-Based Study in Peru. Plos One 6(9): 1-6.

Bertolini DA, Gomes-Gouvêa MS, Carvalho-MelloIM , Saraceni CP, Sitnik R, Grazziotin FG, Laurindo JP, Fagundes NJ, Carrilho FJ, Pinho JR 2012. Hepatitis B virus genotypes from European origin explain the high endemicity found in some areas from southern Brazil. Infect Genet Evol 12(6): 1295-304.

Block TM, Guo H, Guo J 2007. Molecular virology of hepatitis B virus for clinicians. Clin Liver Dis 11(4): 685-706.

Blumberg BS, Alter MJ, Visnich S 1965. A "new" antigen in leukemia sera. JAMA 191: 541-546.

Bond WW, Favero MS, Petersen NJ, Gravelle CR, Ebert JW, Maynard JE 1981. Survival of hepatitis B virus after drying and storage for one week. Lancet 1(8219): 550-551.

Borges AMT, Azevedo MSP, Martins RMB, Carneiro MAS, Naghettini A, Daher RR, Cardoso DDP 1997. Hepatitis B in patients from dialysis centers in Goiânia - Goiás. Rev Pat Trop 26(1): 9-16.

Bossi AP, 2008. The capitalist organization of "informal" work: the case of waste pickers RBCS 23(67): 101-116.

Bottecchia M, Souto FJ, O KM, Amendola M, Brandao CE, Niel C, Gomes SA 2008. Hepatitis B virus genotypes and resistance mutations in patients under long term lamivudine therapy: characterization of genotype G in Brazil. BMC Microbiol 8(11): 1-10.

Bouchard MJ, Schneider RJ 2004. The enigmatic X gene of hepatitis B virus. J Virol 78(23): 12725-12734.

BRAZIL 2005. Ministry of Health. Infectious and parasitic diseases: pocket guide, 6. ed. Brasília, (Series B. Basic Health Texts). 320 p. Available at:

http://bvsms.saude.gov.br/bvs/publicacoes/guia_bolso_5ed2.pdf. Accessed on October 8, 2012.

BRAZIL 2006a. Ministry of Health. Diagnosis of urban solid waste management - 2004. Available at: <http://www.snis.gov.br/>. Accessed on October 8, 2012.

BRAZIL 2006b. Ministry of Health. Health Surveillance Secretariat. Department of Epidemiological Surveillance. Manual of reference centers for special immunobiologicals. Series A. Norms and Technical Manuals. 188p. Available at: http://portal.saude.gov.br/portal/arquivos/pdf/livro_cries_3ed.pdf. Accessed on October 8, 2012.

BRAZIL 2008. Ministry of Health. Health Surveillance Secretariat. Department of Epidemiological Surveillance. Viral hepatitis: Brazil is aware/Ministry of Health, Health Surveillance Secretariat, Epidemiological Surveillance Department. 3ª ed. Brasilia: Ministério da Saúde - (Série B. Textos Básicos de Saúde).   62p. Available at   : http://portal.saude.gov.br/portal/arquivos/pdf/brasil_atento_3web.pdf. Accessed October 10, 2012.

BRAZIL 2010a. Ministry of Health. Health Surveillance Secretariat. Department of Epidemiological Surveillance. Technical note no. 89/2010CGPNIDEVEP/SMS/MS. "Expansion of the supply of hepatitis B vaccine for the 20-29 age group." Available at: http://www.aids.gov.br/legislacao/2012/51047. Accessed on October 10, 2012.

BRAZIL 2010b. Ministry of Health. Health Surveillance Secretariat. Departamento deVigilanciaEpidemiológica   .   Parecertécnicon° 04/2010/CGPNI/DEVEP/SMS/MS and AIDS and VIRAL HEPATITIS/SMS/MS Update on the indication of hepatitis B vaccine in health services. Available at: http://www.aids.gov.br/publicacao/parecer-tecnico-n-042010cgpnidevepsvsms- e-dst-aids-e-hepatites-viraissvsms. Accessed on October 11, 2012.

BRAZIL 2010c. Ministry of Labor and Employment. Secretariat for Public Employment Policies. Brazilian classification of occupations and inter-ministerial committee contribute to the inclusion and recognition of waste pickers. Available at: http://www.setor3.com.br/jsp/default.jsp?tab=00002&subTab=00000&newsID=a43 20.htm&template=58.dwt&testeira=33. Accessed on October 11, 2012.

BRAZIL 2011a. Ministry of Health. Health Surveillance Secretariat. Department of STD, AIDS and Viral Hepatitis. Epidemiological bulletin on viral hepatitis. Year II, n°1, 12-14p. Available at: http://www.aids.gov.br/publicacao/2011/boletim_epidemiologico_hepatites_virais_ 2011. Accessed on October 11, 2012.

BRAZIL 2011b. Ministry of Health. Health Surveillance Secretariat. Department of STD, AIDS and Viral Hepatitis. Clinical protocol and therapeutic guidelines for the treatment of chronic viral hepatitis B and co-infections. Brasília-DF. 13-87 p. Available at: http://portal.saude.gov.br/portal/arquivos/pdf/prot_clinico_diretrizes_terapeuticas_h ep_b.pdf. Accessed on October 12, 2012.

BRAZIL 2012a. Ministry of Health. Health Portal - SUS. Health extends age range for hepatitis   B vaccination   . Available   at: http://portalsaude.saude.gov.br/portalsaude/noticia/3890/162/saude-amplia-faixa- age-range-for-vaccination-of-hepatitis-b.html. Accessed on October 10, 2012.

BRAZIL 2012b. Ministry of Health. Health Surveillance Secretariat. Department of Epidemiological Surveillance. General Coordination of the National Immunization Program. Technical report on the introduction of the pentavalent vaccine - Diphtheria, tetanus, pertussis, hepatitis b (recombinant) and haemophilus influenzae Tipob (conjugate) adsorbed vaccine. Available at: http://www.sgc.goias.gov.br/upload/arquivos/2012-06/informe-tecnico-vacinapentavalent.pdf. Accessed on October 11, 2012.

Bruguera M 2006. Prevention of viral hepatitis. Enferm Infecc Microbiol Clin 24(10): 649-656.

Bruss V 2007. Hepatitis B virus morphogenesis. World J Gastroenterol 13(1): 65-73.

Candotti D, Lin CK, Belkhiri D, Sakuldamrongpanich T, Biswas S, Lin S, Teo D, Ayob Y, Allain JP 2012. Occult hepatitis B infection in blood donors from South East Asia: molecular characterization and potential mechanisms of occurrence. Gut 1-10.

Cardoso DDP, Azevedo MPP, Martins RMB, Barbosa AJ, Camarota SCT 1990. Seroprevalence of hepatitis B virus infection by AgHBs and anti-HBs markers in a female population in an urban area of Goiania-Go. Rev Pat Trop 19(2): 135-141.

Cardoso DDP, Faria EL, Azevedo MSP, Queiroz DAO, Martins RMB, Souza TT, Daher RR, Martelli CMT 1996. Seroepidemiology for hepatitis B virus (HBV) in pregnant women/parturients and its transmission to newborns in Goiania-GO. Rev Soc Bras Med Trop 29(4): 349-353.

Carey WD 2009. The prevalence and natural history of hepatitis B in the 21st[st] century. Cleve Clin J Med 76(3): S2-S5.

Carneiro AF, Daher RR 2003. Seroprevalence of Hepatitis B Virus in Anesthesiologists. Rev Bras Anestesiol 53(5): 672-679.

Carvalho P, Schinoni MI, Andrade J, Vasconcelos Rego MA, Marques P, Meyer R, Araújo A, Menezes T, Oliveira C, Macedo RS, Macedo LS, Leal JC, Matos B, Schaer R, Simones JM, Freire SM, Paraná R 2012. Hepatitis B virus prevalence and vaccination response in health care workers and students at the Federal University of Bahia, Brazil. Ann Hepatol 11(3): 330-7.

Catalan-Soares BC, Almeida RT, Carneiro-Proietti AB 2000. Prevalence of HIV-1/2, HTLV-I/II, hepatitis B virus (HBV), hepatitis C virus (HCV), Treponema pallidum and Trypanosoma cruzi among prison inmates at Manhuagu, Minas Gerais State, Brazil. Rev Soc Bras Med Trop 33(1): 27-30.

Catapreta CAA, Heller L 1999. Association between household solid waste collection and health, Belo Horizonte (MG), Brazil. Rev Panam Salud Publica 5(2): 88-96.

Cavalcante S, Franco MFA 2007. Profession of danger: perception of health risks among waste pickers at the Jangurussu dump. Rev Mal-Estar Subj 7(1): 211-232.

Centers for disease control and prevention 2003. Recommended childhood and adolescent immunization schedule. MMWR Surveill Summ 52(4):   Q1-Q4. Available at   :http://www.cdc.gov/mmwr/preview/mmwrhtml/mm5351-Immunizational.htm. Accessed on October 12, 2012.

Centers for disease control and prevention 2007. Progress in hepatitis B prevention through universal infant vaccination - China, 1997-2006. MMWR Surveill Summ 56(18): 441-445. Available at: http://www.cdc.gov/mmwr/preview/mmwrhtml/mm5618a2.htm. Accessed October 15, 2012.

CDS-UnB/Center for Sustainable Development 2005. "Solid waste is among the emergency problems of future mayors". Available at: http://www.comciencia.br/200412/noticias/2005/lixo.htm. Accessed on October 16, 2012.

Chang JJ, Lewin SR 2007. Immunopathogenesis of hepatitis B virus infection. Immunol Cell Biol 85(1): 16-23.

Chang MH 2007. Hepatitis B virus infection. Semin Fetal Neonatal Med 12(3): 160167.

Chevaliez S, Pawlotsky J.M. 2008. Diagnosis and management of chronic viral hepatitis: Antigens, antibodies and viral and vira gomes. Best Pract Res Clin Gastroenterol 22(6): 1031-1048.

Choo QL, Kuo G, Weiner AJ, Overby LR, Bradley DW, Houghton M 1989. Isolation of a cDNA clone derived from a blood-borne non-A, non-B viral hepatitis genome.

Science 244: 359-362. in J Hepatol 36(5): 582-585, 2002.

Chu CJ, Lok ASF 2002. Clinical significance of hepatitis B virus genotypes. Hepatology 35(5): 1274-1276.

Coffin CS, Lee SS 2009. Treatment of HBeAg-positive patients with nucleos/tide analogues. Liver Int 29 (Suppl 1): 116-24.

Collins CH, Kennedy DA 1992. The microbiological hazards of municipal and clinical wastes. J Appl Bacteriol 73(1): 1-6.

Conjeevaram HS, Lok AS 2001. Occult hepatitis B virus infection: a hidden menace? Hepatology 34(1): 204-6.

Costa EFA 2004. Analysis of seroprevalence for hepatitis B and C virus infections in elderly residents of nursing homes in the municipality of Goiania. Master's thesis - Postgraduate Program in Tropical Medicine - IPTSP/UFG 106p.

Costa CM 2008. Recycling and citizenship: the life trajectory of waste pickers from the Reciclo community. Dissertation (Master's in Education) - Faculty of Education, University of Brasilia, 165p. Available at: http://repositorio.bce.unb.br/bitstream/10482/1889/1/2008_ClaudiaMoraesCosta.pd f. Accessed on October 15, 2012.

Couroucè AM, Holland PV, Muller JY, Soulier JP 1976. HBsAg antigen subtypes. Bibl haemotol (Karger-Basel) n° 42 p.

Couroucè-Pauty AM, Lamaire JM, Roux JF 1978. New hepatitis B surface antigens subtypes inside ad category. Vox Sang 35(5): 304-308.

Couroucé-Pauty AM, Plancon A, Soulier JP 1983. Distribution of HBsAg subtypes in the world. Vox Sang 44(4): 197-211.

Dall'Agnol CM, Fernandes FS 2007. Health and self-care among waste pickers: experiences at work in a recyclable waste cooperative. Rev Latino-Am Enfermagem 15(numero especial): 729-735.

Dane DS, Cameron CH, Briggs M 1970. Virus-like particles in serum of patients with Australia-antigen-associated. Lancet 1(7649): 695-698.

De Franchis R, Meucci G, Vecchi M, Tatarella M, Colombo M, Del Ninno E, Rumi MG, Donato MF, Ronchi G 1993. The natural history of asymptomatic hepatitis B surface antigen carriers. Ann Intern Med 118(3): 191-194.

Dehesa-Violante M, Nunez-Nateras R 2007. Epidemiology of hepatitis virus B and C. Arch Med Res 38(6): 606-611.

Dias AL, Oliveira CM, Castilho Mda C, Silva Mdo S, Braga WS 2012. Molecular characterization of the hepatitis B virus in autochthonous and endogenous populations in the Western Brazilian Amazon. Rev Soc Bras Med Trop 45(1): 9-12.

Diaz RS, Mendonça JS 2006. Measurement of HBV DNA. Braz J Infect Dis 10(Suppl1): 32-34.

Doo CE, Ghany GM 2010. Hepatitis B virology for clinicians. Clin Liver Dis 14(3): 397-408.

EASL (European Association for the Study of the Liver) Jury 2003. EASL International Consensus Conference on Hepatitis B. September 13-14, 2002: Geneva, Switzerland. Consensus statement (short version). J Hepatol 38(4): 533-40.

EASL (European Association for the Study of the Liver) 2009. EASL Clinical Practice Guidelines: Management of chronic hepatitis B. J Hepatol 50: 227-242.

Elgouhari HM, Abu-Rajab Tamimi TI, Carey WD 2008. Hepatitis B virus infection: Understanding its epidemiology, course, and diagnosis. Cleve Clin J Med 75(12): 881-889.

Elgouhari HM, Abu-Rajab Tamimi TI, Carey WD 2009. Hepatitis B: a strategy for evaluation and management. Cleve Clin J Med 76(1): 19-35.

El-Serag HB 2012. Epidemiology of viral hepatitis and hepatocellular carcinoma. Gastroenterology 142(6): 1264-1273.

Etzel A, Shibata GY, Rozman M, Jorge ML, Damas CD, Segurado AA 2001. HTLV-1 and HTLV-2 infections in HIV-infected individuals from Santos, Brazil: seroprevalence and risk factors. J Acquir Immune Defic Syndr 26(2): 185-190.

Fattovich G 2003. Natural history of hepatitis B. J Hepatol 39 (Suppl1): S50-S58.

Feinstone SM, Kapikian AZ, Purceli RH 1973. Hepatitis A: detection by immune electron microscopy of a viruslike antigen associated with acute illness. Science 182(4116): 1026-8.

Feitelson MA, Lee J 2007. Hepatitis B virus integration, fragile sites, and hepatocarcinogenesis. Cancer Lett 252(2): 157-170.

Ferrari C, Missale G, Boni C, Urbani S 2003. Immunopathogenesis of hepatitis B. J Hepatol 39 (Suppl1): S36-S42.

Ferreira MS 2000. Diagnosis and treatment of hepatits B. Rev Soc Bras Med Trop 33(4): 389-400.

Ferreira JA, Anjos LA 2001. Collective and occupational health aspects associated with municipal solid waste management. Cad Saúde Pública 17(3): 689-696.

Ferreira CT, Silveira TR 2004. Viral Hepatitis: epidemiological and preventive aspects. Rev Bras Epidemiol 7(4): 473-487.

Ferreira SL. 2005. The "garbage collectors" in the construction of a new culture: that of separating garbage and environmental awareness. Maringá (PR). Revista Urutágua-revista académicamultidisciplinar n°07. Disponívelem : http://www.urutagua.uem.br/007/07ferreira.htm. Accessed October 16, 2012.

Ferreira RC 2008. Hepatitis b virus infection in illicit drug users in the Midwest region of Brazil: epidemiological and molecular aspects. PhD thesis - IPTSP. Available at: http://posstrictosensu.iptsp.ufg.br/uploads/59/original_TeseRenataCarneiroFerreira. pdf. Accessed on October 16, 2012.

Ferreira RC, Rodrigues FP, Teles SA, Lopes CL, Motta-Castro AR, Novais AC, Souto FJ, Martins RM 2009. Prevalence of hepatitis B virus and risk factors in Brazilian non-injecting drug users. J Med Virol 81(4): 602-9.

Focaccia R 2002. Clinical picture of acute benign forms. In: Veronesi R. & Foccacia R. Tratado de Infectologia. 2 ed, Sao Paulo: Atheneu. 1: 289-290.

Fonseca JCF 2007. Natural history of chronic hepatitis B. Rev Soc Bras Med Trop 40(6): 672-677.

Fonseca JCF 2010. History of viral hepatitis. Rev Soc Bras Med Trop 43(3): 322330.

Franco E, Bagnato B, Marino MG, Meleleo C, Serino L, Zaratti L 2012. Hepatitis B: Epidemiology and prevention in developing countries. World J Hepatol 4(3): 7480.

Fuente RA, Gutiérrez ML, Garcia-Samaniego J, Fernández-Rodriguez C, Lledó JL, Castellano G 2011. Pathogenesis of occult chronic hepatitis B virus infection. World J Gastroenterol 17(12): 1543-8.

Ganem D, Prince AM 2004. Mechanisms of disease: hepatitis B virus infection - natural history and clinical consequences. N Engl J Med 350(11): 1118-1129.

Gilbert RJC, Beales L, Blond D, Simon MN, Lin BY, Chisari FV 2005. Hepatitis B small surface antigen particles are octahedral. Proc Natl Acad Sci USA 102(41): 14783-14788.

Godoy TMP 2005. The Space of Solidarity Production of Waste Pickers - Uses and Contradictions - Dissertation (Master's Degree in Spatial Organization; Geography) - Institute of Geosciences and Exact Sciences, São Paulo State University, 150p. Available at: http://www.ces.uc.pt/nucleos/ncps/ecosol/investigadores/tatiane_godoy/publicacoes /2005_dissertacao_tatiane_godoy.pdf. Accessed on October 15, 2012.

Gonçales Jr FL 2002. Hepatitis B. In: Veronesi R. & Focaccia R. Tratado de Infectologia. 2ª edition. Sao Paulo: Atheneu, vol. 1, chap. 24, p.302-315.

Gonçales NSL, Cavalheiro NP 2006. Serological markers of hepatitis B and their interpretation. Braz J Infect Dis 10(Suppl1): 19-22.

Gonçales NSL, Gonçales Jr FL 2006. Anomalous serologic profiles, genotypes and mutants of HBV. Braz J Infect Dis 10(Suppl1): 23-28.

Gonçalves RS 2004. Waste pickers: life trajectories, work and health. Dissertation (Master's Degree in Public Health) - Sérgio Arouca National School of Public Health, FIOCRUZ - Rio de Janeiro, 96p. Available at: http://teses.icict.fiocruz.br/pdf/goncalvesrsm.pdf. Accessed on October 15 2012.

Gonzales HV, Salinas JL 2009. Natural history of chronic hepatitis B virus infection. Rev Gastroenterol Perú 29(2): 147-157.

Gutberlet J, Baeder AM 2008. Informal recycling and occupational health in Santo André, Brazil. Int J Environ Health Res 18(1): 1-15.

Hatzakis A, Magiorkinis E, Haida C 2006. HBV virological assessment. J Hepatol 44 (Suppl1): S71-76.

Heermann KH, Kruse F, Seifer M, Gerlich WH 1987. Immunogenicity of the gene S and Pre-S domains in hepatitis B virions and HBsAg filaments. Intervirology 28(1): 14-25.

Henkler F, Waseem N, Golding MH, Alison MR, Koshy R 1995. Mutant p53 but not hepatitis B virus X protein is present in hepatitis B virus-related human hepatocellular carcinoma. Cancer Res 55(24): 6084-91.

Hennessey KA, Kim AA, Griffin V, Collins NT, Weinbaum CM, Sabin K 2008. Prevalence of infection with hepatitis B and C viruses and co-infection with HIV in three jails: A Case for Viral Hepatitis Prevention in Jails in the United States. J Urban Health 86(1): 93-105.

Hollinger FB 1996. Hepatitis virus B. In: Knippe, D.M [et al] Fields Virology, Philadelphia: Lippincott-Raven 3 ed. 1(86): 2739-2807.

Hollinger FB, Liang TJ 2001. Hepatitis B virus. In: Knipe, D.M.; Howley, P.M. (eds.): Fields Virology. Fourth ed. Fhiladelphia: Lippincott Willinans & Wilkins 29713036.

Hollinger FB 2007. Hepatitis B virus genetic diversity and its impact on diagnostic assays. J Viral Hepat 14(1): 11-15.

Hollinger FB, Sood G 2010. Occult hepatitis B virus infection: a covert operation. J Viral Hepat 17(1): 1-15.

Hou J, Liu Z, Gu F 2005. Epidemiology and prevention of hepatitis B virus infection. Int J Med Sci 2(1): 50-57.

Hourvitz A, Mosseri R, Solomon A, Yehezkelli Y, Atsmon J, Danon YL, Koren R, Shouval D 1996. Reactogenicity and immunogenicity of a new recombinant hepatitis B vaccine containing Pre S antigens: a preliminary report. J Viral Hepat 3(1): 37-42.

Hübschen JM, Mbah PO, Forbi JC, Otegbayo JA, Olinger CM, Charpentier E, Muller CP 2011. Detection of a new subgenotype of hepatitis B virus genotype A in Cameroon but not in neighboring Nigeria. Clin Microbiol Infect 17(1): 88-94.

Huy TT, Ushijima H, Sata T, Abe K 2006. Genomic characterization of HBV genotype F in Bolivia: genotype F subgenotypes correlate with geographic distribution and T(1858) variant. Arch Virol 151(3): 589-97.

Huy TTT, Sall AA, Reynes JM, Abe K 2008. Complete genomic sequence and phylogenetic relatedness of hepatitis B virus isolates in Cambodia. Virus Genes 36(2): 299-305.

IBGE 2010. Demographic Census: Population and Household Characteristics - Universe Results. Rio de Janeiro 2010. Available at: http://www.ibge.gov.br/home/estatistica/populacao/censo2010/sinopse.pdf. Accessed on October 15, 2012.

ICTV 2011. International Committee on Taxonomy of Viruses. Virus Taxonomy: 2011 Release (current). Hepatitis B virus. In: ICTVdB - The Universal Virus Database, version 4. Büchen-Osmond, C. (Ed), Columbia University, New York, USA. Available at: http://www.ictvonline.org/virusTaxonomy.asp. Accessed on July 19, 2012.

Inchauspé G, Michel ML 2007. Vaccines and immunotherapies against hepatitis B and hepatitis C viruses. J Viral Hepat 14(1): 97-103.

Inoue K, Yoshiba M, Sekiyama K, Okamoto H, Mayumi M 1998. Clinical and molecular virological differences between fulminant hepatic failures following acute and chronic infection with hepatitis B virus. J Med Virol 55(1): 35-41.

Jonas MM 2009. Hepatitis and pregnancy: an underestimated issue. Liver Int 29(1): 133-139.

Joshi N, Kumar A 2001. Immunoprophylaxis of hepatitis B virus infection. Indian J Med Microbiol 19(4): 172-183.

Jung MC, Pape GR 2002. Immunology of hepatitis B infection. Lancet Infect Dis 2(1): 43-50.

Junqueira AL, Tavares VR, Martins RM, Frauzino KV, Silva AM, Rodrigues IM, Minamisava R, Teles SA 2011. Presence of maternal anti-HBs antibodies does not influence hepatitis B vaccine response in Brazilian neonates. Mem Inst Oswaldo Cruz 106(1): 113-6.

Kann M, Bischof A, Gerlich WH 1997. In vitro model for the nuclear transport of the hepadnavirus genome. J Virol 71(2): 1310-1316.

Kao JH, Chen DS 2002. Global control of hepatitis B virus infection. Lancet Infect Dis 2(7): 395-403.

Kao JH 2002. Hepatitis B viral genotypes: clinical relevance and molecular characteristics. J Gastroenterol Hepatol 17(6): 643-650.

Kao JH 2011. Molecular Epidemiology of Hepatitis B Virus. Korean J Intern Med 26(3): 255-261.

Kay A, Zoulin F 2007. Hepatitis virus genetic variability and evolution. Virus Res 127(2): 164-176.

Keramat F, Eini P, Majzoobi MM 2011. Seroprevalence of HIV, HBV and HCV in persons referred to hamadan behavioral counseling center, West of Iran. Iran Red Crescent Med J 13(1):42-6.

Kew MC 1998. Hepatitis viruses and hepatocellular carcinoma. Res Virol 149(5): 257262.

Kew MC 2010. Epidemiology of chronic hepatitis B virus infection, hepatocellular carcinoma, and hepatitis B virus-induced hepatocellular carcinoma. Pathol Biol (Paris) 58(4): 273-277.

Khouri ME, Santos VA 2004. Hepatitis B: Epidemiological, immunological and serological considerations emphasizing mutation. Rev Hosp Clin Fac Med Sao Paulo 59(4): 216-224.

Kidd-Ljunggren K, Couroucé AM, Oberg M, Kidd AH 1994. Genetic conservation within subtypes in the hepatitis B virus pre-S2 region. J Gen Virol 75(Pt 6): 148590.

Kirchner RM, Saidelles APF, Stumm EMF 2009. Perceptions and profile of waste pickers in a city in Rio Grande do Sul. G&DR 5(3): 221-232.

Koff RS 2003. Hepatitis vaccines: recent advances. Int J Parasitol 33(5-6): 517-23.

Kramvis A, Kew MC 1998. Structure and function of the encapsidation signal of hepadnaviridae. J Viral Hepat 5(6): 357-367.

Kramvis A, Kew M, François G 2005. Hepatitis B virus genotypes. Vaccine 23(19): 2409-2423.

Kramvis A, Arakawa K, Yu MC, Nogueira R, Stram DO, Kew MC 2008. Relationship of serological subtype, basic core promoter and precore mutations to genotypes/subgenotypes of hepatitis B virus. J Med Virol 80: 27-46.

Krugman S, Giles JP, Hammond J 1967. Infectious hepatitis: evidence for two distinctive clinical, epidemiological, and immunological types of infection. JAMA

200:365-373. In: Fonseca JCF 2010. History of viral hepatitis. Rev Soc Bras Med Trop 43(3): 322-330.

Krugman S 1989. Hepatitis B: historical aspects. Am J Infect Control 17(3): 165-7.

Kurbanov F, Tanaka Y, Mizokami M 2010. Geographical and genetic diversity of the human hepatitis B virus. Hepatol Res 40(1): 14-30.

Kwon SY, Lee CH 2011. Epidemiology and prevention of hepatitis B virus infection. Korean J Hepatol 17(2): 87-95.

Lacerda R, Stall R, Gravato N, Tellini R, Hudes S, Hearst N 1996. HIV Infection and risk behaviors among male port workers in Santos, Brazil. Am J Public Health 86(8): 1158-1160.

Larrubia JR 2011. Occult hepatitis B virus infection: A complex entity with relevant clinical implications. World J Gastroenterol 17(12): 1529-1530.

Lavanchy D 2004. Hepatitis B virus epidemiology, disease burden, treatment, and current and emerging prevention and control measures. J Viral Hepat 11(2): 97107.

Lavanchy D 2008. Chronic viral hepatitis as a public health issue in the world. Best Pract Res Clin Gastroenterol 22(6): 991-1008.

Le Bouvier GL 1971. The heterogeneity of Australia antigen. J Infect Dis 123(6): 671675.

Le Seyec J, Chouteau P, Cannie I, Guguen-Guillouzo C, Gripon P 1998. Role of the Pre-S2 domain of the large envelope protein in hepatitis B virus assembly and infectivity. J Virol 72(7): 5573-5578.

Lee H, Park W 2010. Public health policy for management of hepatitis B virus Infection: historical review of recommendaions for immunization. Public Health Nurs 27(2): 148-157.

Lewis-Ximenez LL, do O KM, Ginuino CF, Silva JC, Schatzmayr HG, Stuver S, Yoshida CF 2002. Risk factors for hepatitis B virus infection in Rio de Janeiro, Brazil. BMC Public Health 2(26): 1-5.

Levrero M, Pollicino T, Petersen J, Belloni L, Raimondo G, Dandri M 2009. J Hepatol 51(3): 581-592.

Liang TJ 2009. Hepatitis B: The virus and Disease. Hepatology 49(5): S13-S21.

Liaw YF, Chu CM 2009. Hepatitis B virus infection. Lancet 373(9636): 582-592.

Liu CJ, Kao JH 2007. Hepatitis B virus-related hepatocellular carcinoma: epidemiology and pathogenic role of viral factors. J Clin Med Assoc 70(4): 141-145.

Lledó JL, Fernández C, Gutiérrez ML, Ocaña S 2011. Management of occult hepatitis B virus infection: An update for the clinician. World J Gastroenterol 17(12): 15631568.

Lok ASF, McMahon BJ 2001. Chronic hepatitis B. Hepatology 34(6): 1225-1241.

Lok AS 2004. Prevention of hepatitis B virus-related hepatocellular carcinoma. Gastroenterology 127(5 Suppl 1): S303-9.

Lopes CLR, Martins RMB, Teles SA, Silva AS, Maggi PS, Yoshida CFT 2001. Seroepidemiological profile of hepatitis B virus infection in professionals working in hemodialysis units in Goiánia-Goiás, Central Brazil. Rev Soc Bras Med Trop 34(6): 543-548.

Lurman A 1885. Eire icterusepidemie. Berl Klin Wochenschr 22:   20-23, apud: Hollinger FB 1996. Hepatitis virus B. In: Knippe, D.M [et al] Fields Virology, Philadelphia: Lippincott-Raven 3 ed.1(86): 2739-2807.

MacCallum FO 1947. Homologous serum jaundice. Lancet 2: 691-692.

MacCallum FO 1972. Early studies of viral hepatitis. Br Med Bull 28(2): 105-8.

MacMahon BJ 2010. Natural History of Chronic Hepatitis B. Clinc Liver Dis 14: 381396.

Maddrey WC 2001. Hepatitis B: an important public health issue. Clin Lab 47(1-2): 5155.

Magera M 2003. Waste entrepreneurs: a paradox of modernity. Campinas, SP: átomo.

Mahtab M-Al, Rahman S, Khan M, Karim F 2008. Hepatitis B virus genotypes: an overview. Hepatobiliary Pancreat Dis Int 7(5): 457-464.

Marcelim P 2009. Hepatitis B and hepatitis C in 2009. Liv Int 29(suppl1): 1-8.

Martelli CM, Andrade AL, Cardoso DDP, Sousa LC, Almeida e Silva S, Sousa MA, Zicker F 1990. Seroprevalence and risk factors for hepatitis B virus infection by AgHBs and anti-HBs markers in prisoners and first-time blood donors. Rev Saude Publica 24(4): 270-276.

Martins AC 2007. The search for protection for the work of waste pickers: an analysis of the experience of the Waste and Citizenship Institute in Curitiba, PR. Dissertation (Master's Degree) - Federal University of Ponta Grossa, Ponta Grossa. 107p. Available at: tede.uepg.br/tde_busca/arquivo.php?codArquivo=118. Accessed on October 16, 2012.

Maruyama T, Iino S, Koike K, Yasuda K, Milich DR 1993. Serology of acute exacerbation in chronic hepatitis B virus infection. Gastroenterology 105(4): 11411151.

Matos MAD, Reis NR, Koslowski AG, Teles SA, Motta-Castro ARC, Melo FCA, Gomes SA, Martins RMB 2009. Epidemiological study of hepatitis A, B, and C in the largest Afro-Brazilian isolated community. Trans R Soc Trop Med Hyg 103(9): 899-905.

Matos SB, Jesus AL, Pedroza KC, Sodre HR, Ferreira TL, Lima FW 2012. Prevalence of serological markers and risk factors for bloodborne pathogens in Salvador, Bahia state, Brazil. Epidemiol Infect 15: 1-7.

Matsuoka S, Nirei K, Tamura A, Nakamura H, Matsumura H, Oshiro S, Arakawa Y, Yamagami H, Tanaka N, Moriyama M 2008. Influence of occult hepatitis B virus coinfection on the incidence of fibrosis and hepatocellular carcinoma in chron c hepatitis C. Intervirology 51(5): 352-61.

Medeiros LFR, Macêdo KB 2006. Waste pickers: a profession beyond survival? Psicol Soc 18 (2): 62-71.

Mello ES, Alves VAF 2006. Pathological anatomy of hepatitis B. Braz J Infect Dis 10(1): 36-37. Braz J Infect Dis 10(Suppl1): 32-34.

Mello FCA, Souto FJD, Nabuco LC, Villela-Nogueira CA, Coelho HSM, Franz HCF, Saraiva JCP, Virgolino HA, Motta-Castro ARC, Melo MMM, Martins RMB, Gomes AS 2007. Hepatitis B virus genotypes circulating in Brazil: molecular characterization of genotype F isolates. BMC Microbiol 7: 103.

Mendonça JS, Vigani AG 2006. Natural history of acute and chronic hepatitis B. Braz J Infect Dis 10(Suppl1): 15-18.

Milich DR, Jones JE, Hughes JL, Price J, Raney AK, McLachlan A 1990. Is a function of the secreted hepatitis B e antigen to induce immunologic tolerance in utero? Proc Natl Acad Sci USA 87(17): 6599-6603.

Miranda AE, Vargas PRM, Viana MC 2004. Sexual and reproductive health of female inmates in Brazil. Rev Saude Publica 38: 255-260.

Mir-Nasseri MM, MohammadKhani A, Tavakkoli H, Ansari E, Poustchi H 2011. Incarceration is a major risk factor for blood-borne infection among intravenous drug users. Hepat Mon 11(1): 19-22.

Miura, PCO 2004. Becoming a waste picker: a psychosocial analysis. Unpublished master's thesis, Master's in Social Psychology, supervisor Dr. Bader Sawaia, Pontifical Catholic University of Sao Paulo. Sao Paulo, SP. 90p.

Molner JG, Meyer MF 1940. Jaundice in Detroit. Am J Public Health Nations Health 30(5): 509-515.

Morgan HV, Williamson DA 1943. Jaundice following Administration of Human Blood Products. Br Med J 1(4302): 750-3.

Motta-Castro AR, Martins RM, Yoshida CF, Teles SA, Paniago AM, Lima KM, Gomes SA 2005. Hepatitis B virus infection in isolated Afro-Brazilian communities. J Med

Virol 77(2): 188-193.

Motta-Castro AR, Martins RM, Araujo NM, Niel C, Facholi GB, Lago BV, Mello FC, Gomes SA 2008. Molecular epidemiology of hepatitis B virus in an isolated Afro-Brazilian community. Arch Virol 153(12): 2197-205.

Mu SC, Wang GM, Jow GM, Chen BF 2011. Impact of universal vaccination on intrafamilial transmission of hepatitis B virus. J Med Virol 83(5): 783-90.

Mulyanto, Depamede SN, Surayah K, Tsuda F, Ichiyama K, Takahashi M, Okamoto H 2009. A nationwide molecular epidemiological study on hepatitis B virus in Indonesia: identification of two novel subgenotypes, B8 and C7. Arch Virol 154(7): 1047-59.

Mulyanto, Depamede SN, Wahyono A, Jirintai, Nagashima S, Takahashi M, Okamoto H 2011. Analysis of the full-length genomes of novel hepatitis B virus subgenotypes C11 and C12 in Papua, Indonesia. J Med Virol 83(1): 54-64.

Naumann H, Schaefer S, Yoshida CFT, Gaspar AM, Repp M, Gerlich WH 1993. Identification of a new hepatitis B virus (HBV) genotype from Brazil that expresses HBV surface antigen subtype adw4. J Gen Virol 74(8): 1627-1632.

Negro F 2011. Management of chronic hepatitis B: an update. Swiss Med Wkly 141(w13264): 1-7.

Nelwan EJ, Crevel R, Alisjahbana B, Indrati AK, Dwiyana RF, Nuralam N, Pohan HT, Jaya I, Meheus A, Van der Ven A 2010. Human immunodeficiency virus, hepatitis B and hepatitis C in an Indonesian prison: prevalence, risk factors and implications of HIV screening. Trop Med Int Health 15(12): 1491-1498.

Neuveut C, Wei Y, MA Buendia 2010. Mechanisms of HBV-related hepatocarcinogenesis. J Hepatol 52(4): 594-604.

Ngui SL, Hallet R, Teo CG 1999. Natural and iatrogenic variation in hepatitis B virus. Rev Med Virol 9(3): 183-209.

Niel C, Moraes MTB, Gaspar AMC, Yoshida CFT, Gomes AS 1994. Genetic diversity of hepatitis B virus strains isolated in Rio de Janeiro, Brazil. J Med Virol 44(2): 180-186.

Niederhauser C 2011. Reducing the risk of hepatitis B virus transfusion-transmitted infection. J Blood Med 2: 91-102.

Norder H, Hammas B, Lofdahl S, Courouce AM, Magnius LO 1992. Comparison of the amino acid sequences of nine different serotypes of hepatitis B surface antigen and genomic classification of the corresponding hepatitis B virus strains. J Gen Virol 73(5): 1201-1208.

Norder H, Hammas B, Lee SD, Bile K, Courouce AM, Mushahwar IK, Magnius LO 1993. Genetic relatedness of hepatitis B viral strains of diverse geographical origin and natural variations in the primary structure of the surface antigen. J Gen Virol 74(7): 1341-1348.

Norder H, Courouce AM, Magnius LO 1994. Complete genomes, phylogenetic relatedness, and structural proteins of six strains of the hepatitis B virus, four of which represent two new genotypes. Virology 198(2): 489-503.

Norder H, Courouce AM, Coursaget P, Echevarria JM, Lee SD, Mushahwar IK, Robertson BH, Locarnini S, Magnius LO 2004. Genetic diversity of hepatitis B virus strains derived worldwide: genotypes, subgenotypes, and HBsAg subtypes. Intervirol 47: 289-309.

Nwokediuko SC 2011. Chronic hepatitis B: management challenges in resource-poor countries. Hepat Mon 11(10): 786-93.

Ocana S, Casas ML, Buhigas I, Lledo JL 2011. Diagnostic strategy for occult hepatitis B virus infection. World J Gastroenterol 17(12): 1553-1557.

Okamoto H, Tsuda F, Sakugawa H, Sastrosoewignjo RI, Imai M, Miyakawa Y, Mayumi M 1988. Typing hepatitis B virus by homology in nucleotide sequence: comparison of surface antigen subtypes. J Gen Virol 69(Pt10): 2575-2583.

Okochi K, Murakami S 1968. Observations on Australia antigen in Japanese. Vox Sang 15(5): 374-385.

Oliveira SA, Hacker MA, Oliveira ML, Yoshida CF, Telles PR, Bastos FI 2005. A window of opportunity: declining rates of hepatitis B virus infection among injection drug users in Rio de Janeiro, and prospects for targeted hepatitis B vaccination. Rev Panam Salud Publica 18(4-5): 271-7.

Oliveira MD, Martins RMB, Matos MA, Ferreira RC, Dias MA, Carneiro RC, Junqueira AL, Teles SA 2006. Seroepidemiology of hepatitis B virus infection and high rate of response to hepatitis B virus Butang vaccine in adolescents from low income families in Central Brazil. Mem Inst Oswaldo Cruz 101(3): 251-256.

WHO 1977. Word Health Organization. Advances in viral hepatitis. Available at: http://apps.who.int/iris/bitstream/10665/41223/1/WHO_TRS_602.pdf. Accessed on October 10, 2012.

WHO 2010. Word Health Organization. Progress Towards Global. Immunization Goals -2009. Summary presentation of key indicators. Available at: http://www.who.int/immunization_monitoring/data/SlidesGlobalImmunization.pdf. Accessed on October 17, 2012.

WHO2012a   .   Word Health Organization. Hepatitis B. Available at: http://www.who.int/mediacentre/factsheets/fs204/en/. Accessed October 17, 2012.

WHO 2012b. Guidance on prevention of viral hepatitis B and C among people who inject drugs. Available at: http://apps.who.int/iris/bitstream/10665/75357/1/9789241504041_eng.pdf.Acesso on October 16, 2012

Ott JJ, Stevens GA, Groeger J, Wiersma ST 2012. Global epidemiology of hepatitis B virus infection: new estimates of age-specific HBsAg seroprevalence and endemicity. Vaccine 30(12): 2212-9.

Paiva EMM, Tiplle AFV, Silva EP, Cardoso DDP 2008. Serological markers and risk factors related to hepatitis B virus in dentists in the Central West region of Brazil. Braz J Microbiol 39(2): 251-156.

Papatheodoidis GV, Dimou E, Laras A Papadimitropoulos V, Hadziyannis SJ 2002. Course of virologic breakthroughs under long-term lamivudine in HBeAg-negative precore mutant HBV liver disease. Hepatology 36(1): 219-22.

Paraná R, Almeida D 2005. HBV epidemiology in Latin America. J Clin Virol 34(1): S130-S133.

Pawlotsky JM 2003. Hepatitis B virus (HBV) DNA assays (methods and practical use) and viral kinetics. J Hepatol 39(1): S31-S35.

Pawlotsky JM, Dusheiko G, Hatzakis A, Lau D, Lau G, Liang TJ, Locarnini S, Martin P, Richman DD, Zoulim F 2008. Virologic monitoring of hepatitis B virus therapy in clinical trials and practice: recommendations for a standardized approach. Gastroenterology 134(2): 405-405.

Pereira LM, Martelli CM, Merchán-Hamann E, Montarroyos UR, Braga MC, de Lima ML, Cardoso MR, Turchi MD, Costa MA, de Alencar LC, Moreira RC, Figueiredo GM, Ximenes RA, Hepatitis Study Group 2009. Population-based multicentric survey of hepatitis B infection and risk factor differences among three regions in Brazil. Am J Trop Med Hyg 81(2): 240-247.

Pérez V 2007. Viral hepatitis: historical perspectives from the 20th to the 21st century. Arch Med Res 38(6): 593-605.

Perkins JA, MS, MFA, CMI 2002. Medical and Scientific Illustrations. Available at : http://people.rit.edu/japfaa/infectious.html. Accessed on October 17, 2012.

Petrosillo N, Ippolito G, Solforosi L, Varaldo PE, Clementi M, Manzin A 2000. Molecular epidemiology of an outbreak of fulminant hepatitis B. J Clin Microbiol 38(8): 2975-2981.

Pinho AA, Chinaglia M, Lippman SA, Reingold A, Diaz RS, Sucupira MC, Page K, Díaz J 2011. Prevalence and factors associated with HSV-2 and hepatitis B infections among truck drivers crossing the southern Brazilian border. Sex Transm Infect 87(7): 553-9.

Porto SO, Cardoso DDP, Queiróz DA, Rosa H, Andrade AL, Zicker F, Martelli CM 1994. Prevalence and risk factors for HBV infection among street youth in central Brazil. J Adolesc Health 15(7): 577-581.

Porto MFP, Juncá DCM Goncalves RS, Filhote MIF 2004. Garbage, work and health: a case study with waste pickers in a metropolitan landfill in Rio de Janeiro, Brazil. Cad Saúde Pública 20(6): 1503-1514.

Poulsen OM, Breum NO, Ebbehoj N, Hansen AM, Ivens UI, van Lelieveld D, Malmros P, Matthiasen L, Nielsen BH, Nielsen EM et al. 1995. Collection of domestic waste. Review of occupational health problems and their possible causes. Sci Total Environ 170(1-2):1-19.

Goiânia City Hall 2009. Goiânia Selective Collection Program PGCS. Available at: http://www.goiania.go.gov.br/shtml/coletaseletiva/principal.shtml. Accessed on: October 17, 2012.

Prince AM 1968. An antigen detected in the blood during the incubation period of serum hepatitis. Proc Natl Acad Sci USA 60(3): 814-821.

Raimondo G, Pollicino T, Squadrito G 2003. Clinical virology of hepatitis B virus infection. J Hepatol 39(1): S26-S30.

Raimondo G, Pollicino T, Cacciola I, Squadrito G 2007. Occult hepatitis B virus infection. J Hepatol 46(1): 160-170.

Raimondo G, Caccamo G, Filomia R, Pollicino T 2012. Occult HBV infection. Semin Immunopathol 34(4): 1-14.

Ramos JM, Costa e Silva AM, Martins RM, Souto FJ 2011. Prevalence of hepatitis B and C virus infection among leprosy patients in a leprosy-endemic region of central Brazil. Mem Inst Oswaldo Cruz 106(5):632-4.

Rapicetta M, Ferrari C, Levrero M 2002. Viral determinants and host immune responses in the pathogenesis of HBV infection. J Med Virol 67(3): 454-457.

Rehermann B, Lau D, Hoofnagle JH, Chisari FV 1996. Cytotoxic T lymphocyte responsiveness after resolution of chronic hepatitis B virus infection. J Clin Invest 97(7): 1655-1665.

Reuben A 2002. The thin red line. Hepatology 36(3): 770-3.

Reimer J, Lorenzen J, Baetz B, Fisher B, Rehm J, Haasen C, Backmund M 2007. Multiple viral hepatitis in injection drug users and associated risk factors. J Gastroenterol Hepatol 22: 80-85.

Ribeiro H, Besen GR 2007. Panorama of Selective Collection in Brazil: Challenges and Perspectives Based on Three Case Studies. Interfacehs 2(4): 1-18.

Rizzetto M, Canese MG, Aricô S, Crivelli O, Trepo C, Bonino F, Verme G 1977. Immunofluorescence detection of new antigen-antibody system (delta/anti-delta) associated with hepatitis B virus in liver and in serum of HBsAg carriers. Gut 18(12): 997-1003.

Rizzetto M, Zanetti AR 2002. Progress in the prevention and control of viral hepatitis type B: closing remarks. J Med Virol 67(3): 463-6.

Romero M, Madejón A, Fernández-Rodríguez C, García-Samaniego J 2011. Clinical significance of occult hepatitis B virus infection.World J Gastroenterol 17(12): 1549-52.

Rosa H, Costa APVF, Ferraz ML, Pedroza SC, Andrade ANSS, Martelli CMT, Zicker F 1992. Association between leprosy and hepatitis B virus infection. Prevalence study carried out in Goiania, Central Brazil. Rev Inst Med Trop São Paulo 34(5):421-426.

Rozman MA, Alves IS, Porto MA, Gomes PO, Ribeiro NM, Nogueira LA, Caseiro MM, da Silva VA, Massad E, Burattini MN 2007. HIV and related infections in a sample of recyclable waste collectors of Brazil. Int J STD AIDS 18(9): 653-654.

Rozman MA, Alves IS, Porto MA, Gomes PO, Ribeiro NM, Nogueira LA, Caseiro MM, Silva VA, Massad E, Burattini MN 2008. HIV infection and related risk behaviors in a community of recyclable waste collectors of Santos, Brazil. Rev Saude Publica 42(5): 838-843.

Saiz de la Hoya OS, Marco A, García-Guerrero J, Rivera A, Prevalhep study group 2011. Hepatitis C and B prevalence in Spanish prisons. Eur J Clin Microbiol Infect Dis 30(7): 857-862.

Santos AO, Alvarado-Mora MV, Botelho L, Vieira DS, Pinho JRR, Carrilho FJ, Honda ER, Salcedo JM 2010. Characterization of hepatitis B virus (HBV) genotypes in patients from Rondonia, Brazil. Virol J 7: 315.

Scaraveli NG, Passos AM, Voigt AR, Livramento A, Tonial G, Treitinger A, Spada C 2011. Seroprevalence of hepatitis B and hepatitis C markers in adolescents in Southern Brazil. Cad Saude Publica. 27(4):753-8.

Schaefer S 2005. Hepatitis B virus-Significance of genotypes. J Viral Hepat 12(2): 111124.

Seeger C, Mason WS 2000. Hepatitis B virus biology. Microbiol Mol Biol Rev 64(1): 51-68.

Selim HS, Abou-Donia HA, Taha HA, El Azab GI, Bakry AF 2011. Role of occult hepatitis B virus in chronic hepatitis C patients with flare of liver enzymes. Eur J Intern Med 22(2): 187-90.

Seto WK, Lai CL, Yuen MF 2011. Nucleic acid testing for the detection of HBV DNA. Hepat Mon 11(10): 847-8.

Shepard CW, Simard EP, Finelli L, Flore AE, Bell BP 2006. Hepatitis B virus infection: epidemiology and vaccination. Epidemiol Rev 28: 112-125.

Shi Y, WuYH, ZhangWJ, Yang J, Chen Z 2012. Association between occult hepatitis B infection and the risk of hepatocellular carcinoma: a meta-analysis. Liver Int 32(2): 231-40.

Shi YH, Shi CH 2009. Molecular characteristics and stages of chronic hepatitis B virus infection. World J Gastroenterol 15(25): 3099-105.

Shiffman ML 2010. Management of cute hepatitis B. Clin Liver Dis 14(1): 75-91.

Silva CO, Azevedo Mda S, Soares CM, Martins RMB, Ramos CH, Daher RR, Cardoso DDP 2002a. Seroprevalence of hepatitis B virus infection in individuals with clinical evidence of hepatitis in Goiania, Goias: Detection of viral DNA and determination of subtypes. Rev Inst Med Trop Sao Paulo 44(6): 331-334.

Silva, ACG 2002b. Garbage collectors: socio-environmental aspects of the activity developed in the municipal dump of Corumbá, Mato Grosso do Sul. Dissertation (Master's Degree) University of Brasília.

Silva MC, Fassa AG, Siqueira CE, Kriebel D 2005a. World at work: Brazilian ragpickers. Occup Environ Med 62(10): 736-740.

Silva PA, Fiaccadori FS, Borges AM, Silva SA, Daher RR, Martins RMB, Cardoso DDP 2005b. Seroprevalence of hepatitis B virus infection and seroconvertion to anti-HBsAg in laboratory staff in Goiania, Goias. Rev Soc Bras Med Trop 38(2): 153-156.

Silva, PHI 2007. Analysis of reciprocal relations in waste pickers' cooperatives in Brasília. Dissertation (Master's in Sociology), Institute of Social Sciences, University of Brasília, 131p.

Silva JL, de Souza VS, Vilella TA, Domingues AL, Coêlho MR 2011. HBV and HCV serological markers in patients with the hepatosplenic form of mansonic schistosomiasis. Arq Gastroenterol 48(2):124-30.

Simon K, Lingappa VR, Ganem D 1998. Secreted hepatitis B surface antigen polypeptides are derived from a transmembrane precursor. J Cell Biol 107: 21632168.

Siqueira MM, Moraes MS 2009. Collective health, urban solid waste and waste pickers.

Ciênc saúde coletiva 14(6): 2115-2122.

Sitnik R, Sette H Jr, Santana RA, Menezes LC, Graça CH, Dastoli GT, Silbert S, Pinho JR 2007. Hepatitis B virus genotype E detected in Brazil in an African patient who is a frequent traveler. Braz J Med Biol Res 40(12): 1689-92.

Souza MM, Barbosa MA, Borges AM, Daher RR, Martins RM, Cardoso DD 2004. Seroprevalence of hepatitis B virus infection in patients with mental problems. Rev Bras Psiquiatr 26(1): 35-8.

Stief AC, Martins RM, Andrade SM, Pompilio MA, FernandesSM , Murat PG, Mousquer GJ, Teles SA, Camolez GR, Francisco RB, Motta-Castro AR 2010. Seroprevalence of hepatitis B virus infection and associated factors among prison inmates in state of Mato Grosso do Sul, Brazil. Rev Soc Bras Med Trop 43(5): 5125.

Strazza L, Azevedo RS, Heraclito B 2007. Study of behavior associated with HIV and HCV infection in inmates of a prison in Sao Paulo, Brazil. Cad Saude Publica 23(1): 197-205.

Stuyver L, De Gendt S, Van Geyt C, Zoulim F, Fried M, Schinazi RF, Rossau R 2000. A new genotype of hepatitis B virus: complete genome and phylogenetic relatedness. J Gen Virol 81(1): 67-74.

Sutton AJ, Gay NJ, Edmunds WJ, Hope VD, Gill ON, Hickman M 2006. Modelling the force of infection for hepatitis B and hepatitis C in injecting drug users in England and Wales. BMC Infect Dis 6: 93.

Tamura K, Dudley J, Nei M, Kumar S 2007. MEGA 4: Molecular Evolutionary Genetics Analysis (MEGA) Software Version 4.0. Mol Biol Evol 24(8):1596-9.

Tatematsu T, Tanaka Y, Kurbanov F, Sugauchi F, Mano S, Maeshiro T, Nakayoshi T, Wakuta M, Miyakawa Y, Mizokami M 2009. A Genetic Variant of Hepatitis B virus Divergent from Known Human and Ape Genotypes isolated from a Japanese patient and provisionally assigned to new genotype J. J Virol 83(20): 10538-10547.

Tavares RS, Barbosa AP, Teles SA, Carneiro MAS, Lopes CLR, Silva AS, Yoshida CFT, Martins RMB 2004. Hepatitis B virus infection in hemophiliacs in Goias: seroprevalence, associated risk factors and vaccine response. Rev Bras Hematol Hemoter 26(3): 183-188.

Te HS, Jesen DM 2010. Epidemiology of hepatitis B and C viruses: a global overview. Clin Liver Dis 14(1): 1-21.

Teles SA, Martins RMB, Silva AS, Gomes DMF, Cardoso DDP, Vanderborght BO, Yoshida CFT 1998. Hepatitis B virus infection profile in Central Brazilian hemodialysis population. Rev Inst Med Trop Sao Paulo 40(5): 281-286.

Teles SA, Martins RMB, Gomes Sa, Gaspar AM, Araujo NM, Souza KP, Carneiro MA, Yoshida CFT 2002. Hepatitis B virus transmission in Brazilian hemodialysis units: serological and molecular follow-up. J Med Virol 68: 41-49.

Thompson JD, Gibson TJ, Plewniak F, Jeanmougin F, Higgins DG 1997. The Clustal_X windows interface: flexible strategies for multiple sequence aligment aided by quality analysis tools. Nucleic Acids Res 25(24):4876-82.

Tran TT, Trinh TN, Abe K 2008. New complex recombinant genotype of hepatitis B virus identified in Vietnam. J Viral 82(11): 5657-63.

Thedja MD, Muljono DH, Nurainy N, Sukowati CH, Verhoef J, Marzuki S 2011. Ethnogeographical structure of hepatitis B virus genotype distribution in Indonesia and discovery of a new subgenotype, B9. Arch Virol 156(5): 855-68.

Tseng FC, O'Brien TR, Zhang M, Kral AH, Ortiz-Conde BA, Lorvick J, Busch MP, Edlin BR 2007. Seroprevalence of hepatitis C virus and hepatitis B virus among San Franscisco injection drug users, 1998 to 2000. Hepatology 46(3): 666-671.

Valsamakis A 2007. Molecular Testing the Diagnosis and Management of Chronic Hepatitis B. Clin Microbiol Rev 20(3): 426-439.

Van Deursen FJV Hino K, Wyatt D, Molyneaux P, Yates P, Wallace LA, Dow BC,

Carman WF 1998. Use of PCR in resolving diagnostic difficulties potentially caused by genetic variation of hepatitis B virus. J Clin Pathol 51(2): 149-153.
Van Herck K, Vorsters A, Van Damme P 2008. Prevention of viral hepatitis (B and C) reassessed. Best Pract Res Clin Gastroenterol 22(6): 1009-1029.
Velloso MP, Santos EM, Anjos LA 1997. Work process and accidents in household waste collectors in the city of Rio de Janeiro, Brazil. Cad Saude Publica 13(4): 693-700.
Velloso MP, Valadares JC, Santos EM 1998. Household waste collection in the city of Rio de Janeiro: a case study based on workers' perceptions. Ciênc Saude Coletiva 3(2): 143-50.
Vierling JM 2007. The immunology of hepatitis B. Clin Liver Dis 11(4): 727-759.
Wei Y, Neuveut C, Tiollais P, Buendia MA 2010. Molecular biology of the hepatitis B virus and role of the X gene. Pathol Biol 58(4): 267-272.
Wong DK, Huang FY, Lai CL, Poon RT, Seto WK, Fung J, Hung IF, Yuen MF 2011. Occult hepatitis B infection and HBV replicative activity in patients with cryptogenic cause of hepatocellular carcinoma. Hepatology 54(3): 829-36.
Xu XW, Chen YG 2006. Current therapy with nucleoside/nucleotide analogs for patients with chronic hepatitis B. Hepatobiliary Pancreat Dis Int 5(3): 350-358.
Yun-Fan L, Chia-Ming C 2009. Hepatitis B virus infection. Lancet 373: 582-592.
Yu H, Yuan Q, Ge SX, Wang HY, Zhang YL, Chen QR, Zhang J, Chen PJ, Xia NS 2010. Molecular and Phylogenetic Analyses Suggest an Additional Hepatitis B Virus Genotype "I". Plos One 5(2): 92-97.
Zhang Z, Torii N, Hu Z, Jacob J, Liang TJ 2001. X-deficient woodchuck hepatitis virus mutants behave like attenuated viruses and induce protective immunity in vivo. J Clin Invest 108(10): 1523-1531.
Zuckerman AJ 1965. The clinical and laboratory features of acute hepatitis in the Royal Air Force. Mon Bull Minist Health Public Health Lab Serv 24: 340-346.

Buy your books fast and straightforward online - at one of world's fastest growing online book stores! Environmentally sound due to Print-on-Demand technologies.

Buy your books online at
www.morebooks.shop

Kaufen Sie Ihre Bücher schnell und unkompliziert online – auf einer der am schnellsten wachsenden Buchhandelsplattformen weltweit! Dank Print-On-Demand umwelt- und ressourcenschonend produzi ert.

Bücher schneller online kaufen
www.morebooks.shop

Printed by Books on Demand GmbH, Norderstedt / Germany